William Handford Hershman

Manual of Nature Study by Grades

To Accompany the Course of Study for the City and Town Schools of Indiana

William Handford Hershman

Manual of Nature Study by Grades
To Accompany the Course of Study for the City and Town Schools of Indiana

ISBN/EAN: 9783337779726

Printed in Europe, USA, Canada, Australia, Japan

Cover: Foto ©Thomas Meinert / pixelio.de

More available books at **www.hansebooks.com**

MANUAL

OF

NATURE STUDY

BY GRADES

TO ACCOMPANY THE COURSE OF STUDY FOR THE
CITY AND TOWN SCHOOLS OF INDIANA

BY

W. H. HERSHMAN, A. B.

New Albany, Ind.

———————

CHICAGO:

A. FLANAGAN, PUBLISHER.

PREFACE.

This book was written to assist teachers in developing more fully than the limited space of a suggestive course of study would allow, the nature work laid out in the course of study for city and town schools of Indiana. Two objects in the nature study are kept constantly in mind; first, to arouse and cultivate the habit of observation, and second, to impress the facts thus acquired upon the mind. Mr. Hershman recognizes the truth that children have a deep, strong, instinctive love for all things that live and all things that support life.

From a close, personal acquaintance with the author for more than twenty years, I know that this book is the outgrowth of a rich, varied and thoughtful experience with child nature and the nature through which the child lives. It may be said that while this book is written from the scientific side, it is pedagogically correct, and more, it has a freshness of spirit that is in itself one of the most potent factors in education.

> "He gave us eyes to see these,
> And lips that we might tell
> How great is God Almighty,
> Who hath made all things well."

D. M. GREETING.

Indianapolis, Ind., October, 1898.

CONTENTS.

A PLEA FOR THE STUDY OF NATURE.

The purpose in all education is to train the child into a habit of correct thinking; to make him strong to battle with the evils of the world; to lead him to be a good citizen; to perfect him in love for all God's creatures; in short, to enable him to live completely.

Whatever may be my thoughts in regard to the rank of nature study when compared with the educational values of other studies, or whatever may be said in reference to its relation or correlation with them, it cannot be denied that it has great educational value as a factor in the development of human character. This fact is thought to be a sufficient cause for the introduction of such study into the grades.

How nature study serves its purpose in the development of character.

It has been said that every child is born into this world with a two-fold nature, an inner spirit and an outer manifestation. His value throughout life depends upon his relation to the outer world of which he is the center. To him everything appears to be within easy reach. Even the moon and the stars are his playthings; all things are for his use, but it requires effort to bring them under his control. The world is full of life and beauty, ready to contribute to his growth and happiness. The inner spiritual nature of the child is reaching upward to a higher ideal. The ideal just out of reach moves onward and persuades the spirit to follow in pursuit.

To carry intelligence into the vegetable kingdom, the full-grown, well-developed stalk of corn in full ear is the ideal in the life of this plant throughout all stages of its existence. In the realization of this ideal the vital force of this plant uses all its surroundings; soil and moisture beneath, air and carbon dioxide above, and warmth and sunlight all around it. When these outer elements are brought into contact with the inner life of the plant, they are transformed into a thing of beauty and pushed outward as an embodiment and manifestation of the beautiful life within. Throughout the period of growth in the attainment of its ideal, the corn plant is harmonizing the surrounding elements with its own beautiful life.

Or, to carry intelligence into the animal kingdom, the ideal towards which the spirit strives is the full-grown, well-developed animal, and we have only to observe its growth and habits in life to know its place in nature. Here again the crude elements of the outer world are changed into a body of strength, activity and beauty appropriate to the character of the internal spirit. The earth itself acts in obedience to the same laws. It balances the other planets in space and assists in the equilibrium of the whole universe. It receives the elements from the sun and works them over into stormy seas and pacific lakes, rugged mountains and peaceful valleys, majestic forests and grassy prairies, all expressive of that inner life that enables it to fit into the environment—the universe.

As with the examples just given, so with the child. He, too, reaches upward toward an ideal. He is in the world to learn his place in nature that he may adapt himself to his surroundings. He touches nature. He is nature himself and all his acts are nature. The first few years of his life are spent in exploring nature.   He finds systems and plans in nature and his thoughts go out in search of the Great Systematizer and Designer.   This habit of searching and experimenting grows on him until he finds God in nature, and learns to read His thoughts as expressed in the flowers of the field, the trees of the forest and in all other living realities.   The more he reads divine thoughts as expressed in the creation, the more self becomes crucified and the nearer he comes to "Him whose thought nature is." Shall we not, then, give the child the freest opportunity to push upward in the direction of the highest ideal of human character?

The boy feels that there is life force in plant life and intelligence in animal life just the same as in human life, and that the same hand is back of it all; and that the same spirit that developed infinite divisibility and individuality has also brought everything into one grand unity as a manifestation of the universal spirit.   When the child is led to see that life grows out of contrast, and that beauty is found in unified variety, that all nature is formed upon one common plan, and that the same spirit pervades all, he and nature will be blended into one, in which unity they will ever walk, each contributing to the support of the other. Nature flows into the child's life, elevates his esthetic and ethical nature, while he in turn, thus strengthened, contributes to the life of nature and lifts it into grander beauty. Can such experience fail to prepare the child for complete living?

Let us see what the love of nature did for the Greeks and Romans.   They loved and recognized her as their mother.   In fact, they saw in her the workings of the divine spirit.   Their ideas of deity took form, the varieties

of which were as numerous as the ideas to be represented.
Hence wood and stream were early peopled with divine
images born out of this love of nature. "To those gods,"
it is said by one writer, "we owe our grandest architectural
forms and most beautiful statuary. For at first temples
were hollowed out of the trunks of trees, and wooden gods
were placed therein for safety." As the love of nature
lifted man's soul, "temples of wood took the place of trees,
and these in turn gave place to temples of stone, beautifully
adorned with gold and silver, and the wooden gods gave
place to statuary of marble and ivory, so that to-day we can
carve nothing to equal the work of these old Greek sculp-
tors."

The Greek's love of nature developed the Grecian spirit,
and as it grew it poured itself out into the general spirit of
nature, and the spirit of nature, thus reinforced, returned a
flood of light upon the spirit of the Greek. Each stage of
spiritual progress demands a finer piece of statuary to represent deity and a better temple for his dwelling-place. This
idea of worship—for that is what it was—this reaching out
after satisfaction in nature, increases the magnitude, beauty
and grandeur of the statuary and decreases the number of
deities. When each element of the universe was considered
separate and apart from all others, as distinct organisms in
nature, deities multiplied in great abundance; but when the
elements were found organized into one complex whole, a
universal spirit was plainly visible through these outward
manifestations. This universal spirit, which is God in
nature, demanded a temple infinitely more beautiful than
the finest Grecian architecture—a temple not made of material things. The ideal temple moved on and on beyond the
bounds of matter; indeed it passed into the spiritual realm.

The ideal deity also passed beyond marble or ivory, beyond the reach of the sculptor's chisel, and the *Athenian's unknown God made His eternal home in the temple of the* HUMAN HEART.

Did we say awhile ago that we owe our progress in sculpturing and statuary to the ancient gods of Greece? Nay, not so. We owe it all to love of nature and the idea of worship found in nature.

We now see clearly that this temple of God in the human heart should be in absolute harmony with all the grandest products of nature. Within the innermost recesses of this temple the peace of nature mingles with that higher peace, and begets joy and love as heavenly blessings to the human soul. Now tell me if this does not make character, true, beautiful and good!

We go through this world with eyes, but cannot see; ears, but do not hear, for these organs have not been fully opened to the soul. Men of means go yearly to such places of resort as Niagara Falls, Vesuvius and the National Park, and trample under foot daily and hourly many microscopic wonders far more appealing to the soul. Shall we continue so to do; or shall we open up these avenues to the soul, that a flood of light from the outer world may be thrown upon our inner world?

Let us, dear Teachers, endeavor to lead our children so that the primroses and flowers of this earth may pass into the soul, and take root, grow, bloom and throw off fragrance out of the very lives of our children. Such leading will give us a community that will grow in character and happiness, and each individual member thereof will be fully enthroned in all his rights.

W. H. H.

MANUAL OF NATURE STUDY

To Accompany Course of Study for the City and Town Schools of Indiana.

FIRST YEAR.

A.—Plant Life.

1. *Autumn Fruits:* — Peach, Pear, Apple, Grape, etc., as types.

Compare and contrast as to size, shape, color, external coverings, hardness, internal structure, taste, smell. -

Arrangement of seeds, arrangement on stalk, how gathered and marketed, comparison of values, etc.

2. *Autumn Leaves.* — Make collections, study forms, colors, etc.

As types, take leaves of fruits named above and study in connection with the study of those fruits. Make drawings in each case.

3. *Autumn Flowers.* — Goldenrod, Aster and Sunflower.

Encourage the finding of all kinds of Goldenrod in this locality. Make drawings of plant as a whole. Make drawings of leaf, flowers, etc.

4. *Autumn Seeds.*—Make collections.

Study dissemination by winds, animals and currents of water.

As an example of wind dissemination take goldenrod, milk weed pod, thistle, iron weed.

Dissemination by animals may be illustrated by the cockle bur, sand bur, Spanish needle, or other bur-like seeds.

By water, nuts of almost all kinds, linden seed, etc., may serve as types. See if nuts and acorns will float.

5. *Preparation for Winter,* as shown in buds and leaves; make collections of buds:—hickory, buckeye, maple, or fruit trees. Lilac furnishes an example of getting ready for winter. Gather also some buds from house plants, so that children may see that naked buds do not prepare for winter.

6. *Study an Evergreen* as a type form. Compare and contrast with the other trees in regard to shape, size, and color of leaves.

7. *Preparations for Spring.*

a. Planting of seeds in school room,—beans, peas, wheat, oats and corn in earth, sand and water.

b. Observation of germination and growth. These seeds may be planted in cigar boxes, or common earthen flower pots, and watched as far as possible throughout their life history.

8. *Determination of parts of Plants*—root, stem, leaf, bud, flower.

9. *Learn to know Common Flowers.*

B.—ANIMAL LIFE.

1. *Insects.* — Transformation of, collection of cocoons.

Through September, the grasshopper, katy-did, dragon fly, potato beetle, and larvae of butterfly or moth may be observed as to their habitat, mode of eating, life history, etc.

2. *Lessons on Cat, Dog, Horse, Cow, Squirrel, Robin, Blackbird, Wood-pecker and Chicken.* Observe, compare and describe their covering, parts, food, care of young. Illustrate their habits by stories and encourage the children to tell stories about what they have seen.

C.—PHYSIOLOGY.

Learn to name and locate the parts of the body —Head, neck, trunk, arms (right and left), hands, feet. Study movements, use and care of each part; show what can be done by each part ; how adapted to use; kindness, how shown by hands, feet, lips; simple lessons on eating, drinking, breathing, sleeping, with special reference to hygiene and right habits; temperance in eating and drinking.

D.—GEOGRAPHY.

1. *General Position:*—Direction and distance; observation and placing of objects; description by use of prepositions and adjectives.

2. *Particular Position:*—Direction; outdoor observation of the cardinal and semi-cardinal directions.

3. *Forms of Water:*—Cloud, fog, mist, rain, dew, frost, snow, ice; observation of forms as they occur and where they occur, to recognize each and to find the more obvious qualities and uses of each.

4. *Winds.*—Temperature, to recognize by feeling the degrees hot, warm, cold; velocity, to recognize and distinguish by their effects the calm, breeze and gale.

E.—WEATHER STUDY.

Suggestive questions:—Dew, frost, fog, cloudy, clear or partly cloudy. Direction of wind. Kind of night last night. Kind of day.

In connection with this Weather Study, tell the story of Mercury, of Apollo's Cows, Zephyr and His Brothers, The Bag of Winds, Neptune, How Æneas was saved, and Aurora and Her Tears.

In connection with the Plant Study, tell the Story of Clytie, of The Thistle, of Apollo and

Hyacinthus, The Star and the Lily, and The Law of the Wood.

In connection with the Study of Animal Life, tell the Story of Aurora and Tithonus.

Under preparation for winter, tell the Story of the Ant and the Grasshopper, and Broken Wing.

In the Spring-time, tell A Bird Story, The Little Worm that was Glad to be Alive, and Robin Red Breast.

Gems.—1. "Rain Shower," to be given at time of gentle rain. 2. "Little Purple Aster," especially appropriate while studying the Aster. 3. "The Sunbeams," particularly appropriate on the return of a beautiful day after a season of storms. 4. "Leaves at Play," appropriate for a windy day in November. 5. "Sleep, Baby, Sleep," to be recited after a talk about the Moon and Stars. 6. "The Hemlock Tree," to be recited in winter while studying the evergreen. 7. "Catch," from Ben Johnson, a Spring gem.

All these Gems and Stories may be found in Mrs. Wilson's Nature Reader, published by McMillan & Co. Also, see Nature Myths and Stories, by Flora J. Cooke, published by Flanagan.

SECOND YEAR.

A.—PLANT LIFE.

1. *Autumn Fruits.*—Apple, plum, grape, etc. *a.* Collection of. *b.* Study typical forms. *c.* Drawings. *d.* Descriptions, both oral and written. It will require several lessons for the rounding up of the work on fruits.

Review the work of the first year, giving particular attention to comparison in size, color, shape, consistency, external covering, whether hairy or smooth, number of seeds in each, market value of each, etc. Manner of hanging on the tree or vine.

Also discuss to some extent *function* of the fruit, and the dangers through which fruit must pass in order to reach maturity, the adaptability of covering to guard against dangers. These mere hints will prepare the way for a more systematic discussion of the colorations, enemies and protectors, higher up in the grades of school work.

Drawings.—Peach as a whole, apple as a whole, plum as a whole, a single grape, a cluster of grapes; a half peach cut to show the seed, a half apple cut to show the seeds, a half plum cut to show the seed; a peach with a few leaves on twig, an apple with a few leaves on twig, a plum with a few leaves on twig, a bunch of grapes with a few leaves on vine.

2. *Autumn Leaves.*—*a.* Collection of. *b.* Study typical forms. *c.* Drawings. *d.* Descriptions both oral and written.

Before beginning the work on leaves, the teacher should read thoroughly the chapters on leaves in Gray's Botany, or some other good text, and also study the leaves themselves. With Second Year pupils, the external form and appearance are all that can be taught to good advantage. Select several varieties, after teaching the parts of a single leaf, and compare them in regard to size, shape, color, surface, margin and veins. Such exercises will lead to the following conclusions:

1.   There are two kinds of leaves, simple and compound. *a.* Simple leaves have but one blade on a foot stalk. *b.* A compound leaf has two or more bladelets, each usually with a separate petiole, but all joined to one common petiole.

2.   The under surface of leaves is usually lighter in color than the upper surface.

3.   All leaves have veins which proceed from the petiole, but they are arranged in different ways in different leaves.

4.   The margins or edges of leaves are either smooth or cut and notched in various ways.

5.   Leaves vary in shape, size and color, so that the leaf of one kind of plant can always be distinguished from that of another.

For mounting and preserving leaves, the teacher is referred to Howe's Systematic Science Teaching, page 122, D. Appleton & Company.

Drawings.—1. An apple leaf with its petiole and venation. 2. A peach leaf with petiole and venation. 3. A grape leaf with petiole and venation. 4. Sycamore leaf. 5. Maple leaf. 6. Oak leaves of several kinds. 7. A group of buckeye leaves from one bud. 8. A walnut leaf. 9. "A Heaven Tree" leaf, and, 10. Locust leaves.

NOTE.—The teacher may make drawing upon the board to illustrate method of representation, but in no case should the pupils be permitted to draw from a copy. The drawing by the teacher should be immediately erased and the attention of the pupils be directed to the leaf itself.

3. *Autumn Flowers.*—Gentian, golden rod, aster, Jamestown or "jimson weed," sunflower, and thistle.

a. Collection of. *b.* Study typical forms. *c.* Drawings. *d.* Descriptions both oral and written.

After a comparative review of work suggested in the first year, the following facts should be established as far as possible in regard to each plant:

1. Where found, whether in cultivated fields and in heaps of rubbish and rich places as in the case of jimson weed, or along the country road side as in the case of the golden rod, in dry pas-

tures as with the thistles, etc. 2. Its nativity. 3. How it came to this country. 4. Its relations; for instance, the jimson is a near relative of the tobacco plant. 5. Visitors, such as bees, ants, butterflies, flies, etc., and why they go there.

4. *Autumn Seeds.*—*a.* Collections of. *b.* Study of typical forms. *c.* Drawings. *d.* Descriptions both oral and written, but principally oral.

Make collection of acorns, walnuts, hickory nuts, hazel nuts, chestnuts, all with the pod or shuck, if possible, so as to lead to a simple discussion of protection. A cocoanut within the shuck is very interesting by way of comparison with other nuts.

Select, also, beans, corn, oats, always calling attention to the covering which may be compared in each case with the covering of apples, peaches, etc., of the preceding month.

Oral lessons on gathering nuts, corn, oats, beans and cocoanuts will be very valuable to cultivate power of conversation. Comparative values in market may be considered.

5. *Preparation of Plants for Winter.*—As shown by changes in leaves, buds and bark. This topic may be discussed at the conclusion of several lessons on hibernation of animals, which see. Collect a great many buds after the frost has taken off the

leaves and notice the scaly covering in each case.
Compare with naked buds in green house. Notice
the sticky substance that holds these buds together. Compare with naked buds. Bring up
this question again in spring time when the buds
begin to open.

Compare the bark of plants that endure the
winter, with that of green house plants. Why the
difference? Notice difference between bark of this
year's growth and that of last year's growth. How
does it differ at frosting time from the growing
time?

Notice that some plants, like the violet, spring
beauty and potato, go into winter quarters under
the ground just as some animals do.

Observe that trees burst their bark when they
get too large for it, thus making the outside very
rough, as seen in the walnut or bur-oak. How
about the locust or grasshopper, crayfish and cicada
when they grow too large for their skins? The
snake, frog, boy?

6. *Effects of Frost:*—On leaves, buds, stems
and flowers. The suggestions under (5) are ap-
plicable here.

Bring out the thought that some plants die
down to the ground every year when frost comes,
while others only drop their leaves.

7. *Preparation for Spring.*—*a.* Germination of seeds planted in school room. Keep record of frequent observations of growing plants. *b.* Germination and growth of self-sown seeds, maple, acorns, etc. *c.* Flow of sap, growth of stems, leaves and flowers.

8. Study spring flowers as to form and colors. Names of common flowers such as violet, spring beauty, hepatica, mustard, windflower, or anemone, butter-cup, etc. Compare and contrast the humble and modest violet of this spring with the haughty jimson of last fall. See Bryant's poem on the violet.

B.—ANIMAL LIFE.

1. *Insects:*—Ant, bee, beetle, grasshopper, etc. *a.* Collections. *b.* Study typical forms. *c.* Drawings and descriptions both oral and written. *d.* Habits. *e.* Transformation.

2. Covering of animals for the seasons.

3. Habits of hibernation.

4. Prehension of food.

    a. Organs of. *b.* Method of different animals.

5. Reappearance of birds. Notice the instincts shown in migration, nesting and care of young.

6. Study of Tadpole and Frog.

Suggestions on the course:

Collect an ant, a bee, a wasp, a butterfly, a grasshopper, and notice that they are all alike in that
they are all cut into in two places, hence name insect, *cut into.*

1.   Let the pupils point out the head, chest, and
abdomen in each case.

2.   Point out the apparatus that belong to the
head, viz.: mouth, eyes and feelers.

If you have a hand lens pass it around and let
the children see that each insect is provided with
compound eyes so that it can see in every direction without turning the head.

How different is the grasshopper's eye from the
eye of the little boy or cat.   Besides the compound
eyes, one on each side with many little faces, the
insect also has three little, simple eyes.   See if the
children, by aid of the lens, can find them.   Why
do you suppose they have those three simple eyes?

2.   The chest has three pairs of legs below and
two pairs of wings above.   Let the children find
these and count them and see that they all belong
to the chest or thorax.   To little children there
may appear exceptions to this rule as every insect
of the group *diptera*, of which the common house
fly is an example, has but one pair of wings.   In
place of the second pair two knobbed threads, or
"balancers" appear.   But this should not be dis-

cussed in this grade. Also some butterflies have abortive two front legs, thus leaving but two pairs of real legs.

3. Point out the organs of the abdomen if any are to be seen; for example, the sting of the bee, the boring machine called ovipositors in female grasshoppers, cicada, etc., for making holes in which to deposit eggs.

Most insects have nine segments in the abdomen, one in head and three in the thorax. Have the children count to see if that statement is true. Here, again, it may be necessary to use the hand lens, with some insects.

When studying the beetle, have the pupils observe that the outer pair of wings is hard and used as a covering for the delicate, gauzy wings and thus protects them from dirt and other rough things. It will also be observed that some insects have no wings. Ants cut off their own wings when they have no more use for them.

Touch upon the uses of insects. Let the children tell stories about ants and bees as to their way of working. Compare the wisdom and industry of these two insects with the idleness and wastefulness of grasshoppers and crickets.

Observe the transformation of carterpillars. Make drawings on a large scale of each insect studied.

Where do the insects spend the winter? What is the preparation of each for winter?

NOTE:—The queen hornet may be found under the bark of old stumps or trees, the queen bumble bee in some protected corner, under boards, the beetle under rocks, in old logs and rotten stumps; the remainder of the bumble bees, grasshoppers and butterflies die.

What preparation is made by the Squirrel? The Blue Jays? The Robin? The English Sparrow? For information in regard to our Indiana birds see Blatchley's Geological Report of Indiana for 1897.

1.—PHYSIOLOGY.
For December, January and February.
1.

Simple Lessons on the senses and what we learn through them.

a. For example, the eye:—

Location, number, shape of pupil, number of lids, how the lids move, how they are kept moist. Compare with same organ in grasshopper, in chicken, in cat, in cow.

b. *Touch.*—Try touching an object with back of hand, with forehead, with cheek, with tongue, with finger tips.

How does a horse feel? A cow? A cat? A grasshopper? Which way is the best?

c. Hearing.—Compare the outer ear of children with that of horse, rabbit, dog, cat, and determine which is most sensitive. Is it necessary to keep the ear clean? Does it ever injure the ear to take cold? Is the wax of any use to the ear? Are the hairs in the ear of any use? Look in the horse's ear for hairs. In the dog's, the cat's. Where is the bird's ear? What do we learn by means of our ears? By means of our eyes? Our fingers?

d. Smell.—Name of organ, nostrils. Compare with the same in horse, in dog, in cat, in hen, and determine which is the most sensitive. What pleasure do we get from the sense of smell? Why should the organ of smell be placed so close to the mouth? Will a cold in the head injure the sense of smell? How? Should we breathe through the nose rather than the mouth? Why? Would a severe cold prevent us from breathing through the nose? Draw the conclusion that it is best to avoid taking cold.

e. Taste may be dealt with in same way. Discuss relative values of these senses, as to which is most important, which least, etc.

2.—THE SKIN.

Review the skin or rind of the apple, the peach, the plum, and grapes as to protection. Also the bark of the tree as to its purpose, as to its cracking open as the growth of the tree proceeds. From the foregoing facts reach the conclusion that our skin protects inner parts from injury: that our skin is scaly, and these scales become loosened as our body grows, and, instead of cracking open like the bark on the tree, they must be washed off so as to give room for the new skin to form and do its proper work. Is the skin of the same degree of thickness all over the body? Is it just the same on the soles of the feet and the palms of the hands as it is on the other parts of the feet and hands? Can you increase the thickness of any part of the skin? How?

Why has the hickory bud such heavy scales in cold weather, and the geranium bud none? Why is our skin thicker where most exposed? How is the skin kept moist? How does the moisture get through the skin?

Other matter that would be an injury to us, if left in the system, comes out with the moisture. Bring out the thought that there are thousands of these little openings on the skin that must be kept clean, hence the bath and clean under-clothing may be discussed here.

The following conclusions in regard to the skin should be fixed upon the minds of the children:

1.   The skin serves for protection of the body.

It helps to warm us when we are cold, and to cool us when we are warm. It helps to purify the blood.

2. There are tiny tubes or pores that open on the surface of the skin. These tubes permit the sweat and other impurities of the blood to escape.

3. Bathing is necessary in order to keep these pores open so that impurities may continue to escape through them.

4. In the morning, just after rising from bed, is the best time to bathe.

5. After bathing, the skin should be dried well and rubbed with the hands until a warm glow sets in.

6. Cool water is better than warm for bathing purposes.

7. We should not bathe in a room in which the air is so cold as to chill the skin.

8. On retiring at night, all clothing worn during the day should be removed from the body and a clean gown put on to sleep in.

9. Our under-clothing absorbs the impurities as they escape from the pores of the skin, hence it is very filthy to wear under-garments more than a week without change.

Cleanliness of nails and scalp should be urged.

3.—LESSONS ON TEETH.

1. Have the children examine the cat's teeth at home. *a.* As to number. *b.* As to kind, whether long or short, blunt or sharp. *c.* As to use, how the cat eats its food. *d.* Are the cat's teeth filthy and decayed? Do they ever need to be filled? Does the cat ever have toothache?

2. For another lesson get reports from the children in regard to the dog's teeth, comparing every point with that of the cat.

3. A third conversation may be based upon the horse's teeth. His manner of eating. What he eats, different kinds of teeth, color, etc. Tell the children to throw a piece of meat to the dog, a handful of oats to the horse and an ear of corn to the cow and watch them eat. Are the teeth fitted in each case for the work they have to do? Why do they not feed oats to the dog, and meat to the horse? Do horses, cows and dogs often have the toothache? I wonder why.

4. For a fourth conversation the children may now be interested in their own teeth. *a.* As to number. *b.* As to kind, temporary and permanent. *c.* How many are like the dog's teeth? How many like the horse's teeth? Then can a boy bite like a dog and also grind or chew like a horse? What food can a boy eat? Will a dog eat every kind of

food that a boy will eat? Will a horse? Why? Will a horse leave his oats to drink cold water, or hot coffee or tea? Will a boy? Will a dog eat hot potatoes and hot biscuits, and take a drink of cold water to cool the burning mouth and stomach? Will a boy do that way? Will a cat or dog leave between his or her teeth particles of food to rot and smell bad? Will some boys do that way? Do cats and dogs have toothache? Do boys? Grown up boys?

Conclusion to be reached:

1.   Teeth are used for biting and chewing food.

2.   Sudden changes of temperature crack the enamel and thus cause decay.

3.   The use of tobacco is an injury to the teeth.

4.   Decaying particles of food generate an acid that causes the teeth to decay.

5.   The teeth must be thoroughly cleansed after each meal and immediately before retiring at night.

6.   In cleaning teeth a good tooth brush, or flannel cloth, with clean water of same temperature of mouth, should be used.

4.—LESSONS ON THE BONY FRAMEWORK OF THE BODY.

Introduce this subject by a lesson on the umbrella, and let the children discuss freely the use of the stays and ribs.. As analagous topics, the children will perhaps suggest the tent, with its pieces of timbers, center-pole, corner-poles, etc., for support. Almost any of the boys can describe a kite and tell what the frame work is for, and how it is made. From these simple illustrations, a chicken or other animal with which the children are well acquainted, may be introduced and bones examined. Develop the thought that the leg bones support the body and asssist in locomotion; that the wing bones assist in flying; that the vertebrae or neck bones give elasticity to the movements of the head in gathering food or in drinking; that the ribs and back protect the heart, lungs and other vital organs as well as to give shape to the chicken. Strip the flesh from the drum stick and call attention to the slender shaft and enlarged extremities and let the children draw conclusions as to purpose in such arrangement. Notice the smoothness of the joints. Notice that the bone is hollow and filled with marrow.

Imagine a chicken trying to make its form more beautiful by wearing a tight fitting band around its ribs. What would the other chickens think of such a foolish notion? Could a young chicken cock crow as loudly if he were so bandaged? Could he breathe as well? Do you think a young hen so distorted could be recommended to preside over a young brood of chickens? Would Madam Pussy look more handsome if she were bandaged around the ribs and abdomen? Would her heart have as much room to pump blood through the arteries and veins? Could she breathe as well? Would you like to have such a foolish kitty in your house? What sort of a coat does a dog wear? Did you ever feel how loose it is? I wonder why he does not wear tight-fitting clothes? The bones of boys and girls may now be discussed in comparison with the foregoing thoughts, putting emphasis upon the fact that the bones are to give shape, support and protection, and to do these three things to best advantage, they must not be hindered in any way by tight lacing or bandaging the head or feet; also that improper position while sitting or standing is as pernicious as bandaging, especially in growing children. Trees grow crooked if they do not assume the proper position when young. Then will not children do the same?

5.—How Tobacco and Alcohol Affect the Growth of the Bones.

This is difficult to teach in this grade. All that can be done is to lead the children to see that a good house must have the very best of material. It must have a good foundation and good timber for a frame work. That good timber cannot grow in

improper soil or improper air or poisoned water.
That plants grown in one soil and atmosphere will
develop into stronger plants than if grown in another. Let the pupils give instances. Since the
plant which grows in the garden or the timber
which makes the frame work of the house must
have just the right kind of food and drink, so
must the bone of the growing boy have pure blood
in order to make strong bone in manhood. Alcohol and tobacco poison the blood and therefore
dwarf the growth of bone.

Breathing. — *a.* Organs. *b.* Process of. *c.*
Purposes. *d.* Position of body in. *e.* Bad effects of
improper position of body, breathing impure air,
improper breathing.

LESSONS.

1. Begin this subject by a breathing exercise
by the whole school. Require the children to take
a full breath, by standing erect and closing the
mouth. What change is made in form of chest
and abdomen while taking a full inspiration?
During expiration? What muscles are used most
in breathing? Could these muscles act freely if
tightly bandaged? Why do we breathe? That
which receives the air in inspiration is called the
lungs. These lungs surround the heart, which is
the blood pump.

Now what happens to the air in the lungs when they are pressed together by the ribs? What happens to the blood pump? What effect would that have upon the supply of blood to the system? Why do we need the blood? Why do we need air?

Teachers must explain that the blood must be purified and that our breathing helps to do that thing. That there are blood tubes and air passages, capillaries and millions of air cells in the lungs, and that the blood trades off its impurities for the oxygen of the air. The lungs may be compared to a market house in that it is a place for trading. Which makes the best trade, the blood or the air?

When does a gardner find it hard to get rid of his cabbage or radishes? When the market house is already full of such vegetables. When does the blood find it difficult to get rid of its impurities? When the air in the lungs is already full of impurities.

Why is plenty of pure air then necessary? How does stooping over while sitting at the desk affect the amount of pure air in the lungs? How does smoking cigarettes or tobacco of any kind affect the purity of air we breathe? How do close, ill-ventilated school rooms affect the purity of air? Then how would that affect the purity of the blood?

How does the condition of blood affect our health?
Then would it be good for our health to remain in
a privy or water closet too long? To sleep in a
very close room? To breathe tobacco smoke? To
lace the ribs and abdomen? Why?

D.—GEOGRAPHY.

Continue work of first year. Study of forms of
solids, as cube, pyramids, sphere, etc. From the
study of the sphere develop the shape of the earth.
From conversations based upon readings from
"Seven Little Sisters," and similar books, develop
as far as practicable at this stage, an idea of the
earth as a whole.

E.—WEATHER STUDY.

Observations and morning notes. *a.* Dew,
frost, or neither. *b.* Rain or snow. *c.* Direction
of wind at 8 o'clock, a. m. *d.* Clear, cloudy, partly
cloudy, raining or snowing at 8:00 a. m. See direction for work in first grade.

Literature.—See Wilson's Nature Reader.

1. Stories to be told to the children. *a.* "Clytie,"
to be told in connection with the study of the sunflower. *b.* "Story of the thistle," to be told in connection with the study of the thistle. *c.* "Aurora."
d. "Aurora's Tears." *e.* "Aurora and Tithonus,"
to be read in connection with the study of the

grasshopper. *f.* "Apollo's Mother," to be read in connection with the study of the frog. *g.* "Birth of Apollo." *h.* "The Killing of the Python." *i.* Apollo and Hyacinthus, Story of Phaethon. Iris, Diana, Diana and Endymion. The Hottentot Moon Story. Yum Sing. German Story of the Moon. The Ant and the Grasshopper, to be read in connection with the study of these insects. The Kind Old Oak. Callisto and Arcas, or the White Bear. Story of the Peacock.

Spring Stories:—1. Pluto. 2. Proserpine. 3. The Finding of Proserpine. 4. Robin Red Breast. 5. How the Bee got the Sting. 6. Legend of the Spring Beauty. 7. Story of the Poplar.

GEMS TO BE LEARNED.

1. *Autumn Gems.*—Rain Shower, Wind Song, Little Purple Aster, The Sunbeams, Leaves at Play.

2. *Winter.*—Chickadee, Snowflakes, The Hemlock Tree, The New Year Song, Song of the Wrens.

3. *Spring Gems.*—Spring, Celia Thaxter, All the Birds have Come Again, Swallow, Calling the Violet, Dandelion Fashions.

THIRD YEAR.

A.—PLANT LIFE.

1. *Lessons on Leaves.*—*a.* Compare as to form, margin, variation, surface. Take a number of large leaves, as sycamore, maple, linden, catalpa, iron weed, thistle, morning glory, pond lilies, beans, grass, etc. First classify on basis of general outline, viz.:— Linear, lanceolate, wedge-shaped, spatulate, ovate, obovate, kidney-shaped, orbicular, elliptical. Keep repeating this exercise in several lessons until the children are fairly well acquainted with the common forms. Let them find that the forms lanceolate, kidney form, heart shape and ovate are widest near the base and taper toward the apex; and that the forms spatulate, wedge-shape and obovate are widest near the apex and taper toward the base, while the forms orbicular, oblong, elliptical, linear and oval are widest at the middle and taper equally toward apex and base. Make drawings of leaves, showing the several forms. As to margin compare the leaves of cherry, peach, grape, apple, elm, beach, hickory, clover, bean, violet, goldenrod, aster, oak, maple, etc. Which of these leaves have entire edges? Which serrate? Dentate? Crenate? Sinuate, etc? For meaning of these terms, the teacher may consult any good botany.

As to division of blade, compare oak, maple, passion vine, water melon, tongue grass, and learn the meaning of lobed, cleft, parted and divided. As to variation, a number of leaves should be used, representing those with petioles, without petioles; with stipules, without stipules; with bud in axil, as common, simple leaf of almost any sort; without bud in axil, as observed in any leaflet of compound leaf; colors, red, yellow, green, variegated, etc.; venation, parallel veined, netted veined. Draw several kinds of simple leaves, showing venation, petiole, etc. Draw several compound leaves, such as horse chestnut, walnut, locust, acacia, strawberry.

Lead the children to see that the bud of a compound leaf is found in the axil of the entire leaf-stalk and not in the axils of the leaflets. Most simple leaves have buds in the axil. The sycamore and a few others may seem like exceptions to the rule, as the bud in such trees is always under the cup of the leaf-stalk.

Under variation, may also be discussed bud scales, as in hickory or lilac, modified for protection of the green leaves within; the tendrils as in pea vines; spines as in barbery; bulb scales as in onion. This might be carried much further, but the foregoing is thought to be sufficient for third

grade work.   Leaf movements can be studied with
considerable interest.   Have the children observe
the leaves of the locust in day time, and again at
night when the leaves are asleep, and report
change.   Observe also oxalis, clover and acacia in
same way.   Observe that the leaves of these plants
wake up at sunrise ready for their day's labor.
For a discussion of plant movements see Chapter
VI of Caroline A. Greevey's Recreations in Botany, published by Harper & Brothers.   Also her
chapter on leaves is well worth reading.

McBride's Lessons in Botany, by Allyn, Bacon
& Co., and Elements of Botany, by Bergen, Ginn
& Co., will be found very helpful to the teacher.

2. *Lessons on Flowers.*—*a.* Study calyx, sepals,
corolla, petals, stamens, pistils.   *b.* Draw and describe each.   It will be better to postpone this part
of the work until the Spring flowers are in bloom,
as so many of the Autumn flowers belong to the
Compositæ, and, on that account, the teaching of
the parts of a flower would be very difficult and
confusing.   Take, for example, a number of Spring
Beauties, equal to the number of children in the
room.   Tell the pupils that the first, or outer covering, is called the calyx.   Ask the children its
color, so as to be sure they have the right coat in
mind.   Then inquire about the number of parts in

that coat, and tell them that each part is called a
sepal. Now, how many sepals in the calyx?
What are sepals? What is the calyx? What is
its color? The next coat or covering is called the
corolla. What is the color of the corolla? Each
part of the corolla is called a petal. How many
petals in the corolla of the Spring Beauty? What
are petals? What is a corolla? Is a petal like a
sepal? What is the difference between a corolla
and a calyx as to order of covering? As to color?
- As to number of parts? As to name of parts? Is
the calyx smooth or rough? How is the inside of
calyx as compared with outside in regard to smoothness? How about the corolla? Does a sepal look
anything like a leaf? Does a petal? Which looks
the more like a leaf? In what way? Do you
find anything like veining in either of these parts
as you do in leaves? Look closely with hand
lens.

Look for a whorl of several stalks with little,
yellow pods on the end. These stalks are called
stamens. How many stamens are there in the
Spring Beauty? Are they all of the same size?
How are they arranged with reference to the petals?
(Opposite.) How does their number correspond
with the number of petals? With the number of
sepals? You will observe a little, oblong case at

the top of each stamen, that is called the anther.
How many anthers are there? The anther is filled
with a yellow powder, or very small grains, called
pollen. Which way does the anther in the Spring
Beauty open?

The little thread-like stalks that support the
anthers are called filaments. How many filaments
are there? How many anthers? What are anthers and filaments together called? What are the
petals together called? Which whorl is called the
calyx? Which the corolla? Which the stamens?
How many parts in each whorl? What are the
parts in each whorl called?

There is now one thing more for us to learn, and
that is, the little stalk, which is large a the bottom,
slender in the middle and divided into three parts
at the top. This stalk is called the pistil. It
looks like the pestle that the druggist uses to mix
medicine with. Perhaps that is the reason it is
called pistil. (From Latin Pistillum, which means
a pestle. The word first came into use for such
flowers as those of Fritillaria Imperialis, or crown
imperial, where the pistil resembles a pestle and
the perianth around it a mortar of an apothecary.)
The large part at bottom of flower cup is called
ovary, or little seed case, the slender part above is
called the style, the three parts at top are called

the stigma. The little grains in the ovary are called ovules. The teacher should now review all the whorls, and parts of each whorl.

Next ask certain pupils of the class to bring to the school-room on another day some member of the Mustard family, or some other simple flower, so that the children can compare the parts of the new flower with those of the Spring Beauty in every particular. At another time, a different flower, and so on, until a half-dozen flowers have been analyzed and names learned.

Up to this time, nothing has been said about the *function* of the parts of the flower. We may now begin this subject by introducing factories in general. A factory is a building in which goods are manufactured, as a mill, where flour and meal are made. Let the children give other examples, naming in each case the raw material out of which must come the manufactured product. The flower of the Spring Beauty is a factory. It manufactures Spring Beauty seeds. The raw material is pollen and ovules. These two kinds of stuff must be mixed together at just the right time or the goods will be spoiled. The ovary is the hopper, the style is the tube through which the pollen must grow downward to the hopper, where the ovules are. The stigma must be moist when the pollen is let

fall, otherwise it will not grow down the tube. How
careful those little men (the stamens) that hold the
pollen are! They certainly do not sleep much,
for, just as soon as they know the stigma is moist,
they must open their little baskets and let the pollen fall on it. They must be quick in their work,
for the stigma will not stay moist long. Like most
other factories, the Spring Beauty closes up at
night, by drawing its petals and sepals tightly over
the inner parts of the factory, so as to shut out the
night air, rain, and night insects that come to do
harm. This factory, in course of a month or two,
will turn out some fine seed, which will fall to the
ground when the pod splits open, and be ready to
start a new plant for another year. Watch these
Spring Beauties and Mustards to see what they
will do. The old plant will die down when frost
comes, but it, too, will spring up again at the return of the song birds from the South.

Deal with the mustard factory in same way.
Also a few of the others studied. Watch for the
seeds as the pods ripen. See (4) of this course.

Make drawings of spring beauty, mustard, buttercup, bloodroot, or other plants studied.

A Few Familiar Flowers by Margaret Warner
Morley, published by Ginn & Co., is excellent for
teacher's use in teaching any kind of flower.

3. *Lessons on Trees.—a.* Distinguished by leaf, shape, bark, habit. *b.* Observe development of leaf-buds. *c.* Study arrangement of leaves.

The autumn months any time before frost will be the proper time to introduce the study of trees. A trip to the woods by the whole school is the best way to secure the right kind of interest. This may be done immediately after a study of the leaves in the school room, as given in the first part of this outline. The leaves of the sycamore, linden, catalpa, cherry, beech, hickory, oak, elm, horse chestnut, walnut and locust have already been studied as to form, variation, color, etc. So it will be an easy matter now to select the trees that bear them. Before starting on this visit to the woods it will be well for the teacher to supply each of the children with a number of cards on each of which may be written near the upper margin the name of a tree whose leaf has already been studied by the pupil. On entering the woods the children holding these cards will separately search for the tree or trees named on these cards. On finding the object of their search they will write on their respective cards a brief description of bark, and shape of tree, write signature, pin the card upon the tree described and return to the teacher. These cards thus pinned to the tree will stand as evidence of

ownership by discovery, providing no mistake has
been made. The teacher will now go with the pupils, visit all the trees and note the good hits as
well as the errors.

Again, each pupil may be supplied with a few
pieces of bark, with direction to place walnut bark
with a walnut tree, hickory bark with a hickory tree,
and so on, each kind of bark with its respective kind
of tree. As before, the pupil should leave her name
on a piece of paper pinned to the bark so that the
teacher may know who discovered the tree. Test
again with nuts, acorns and other fruits to any extent the teacher may desire, or until the interest
begins to lag. To vary the program let the children gather a miscellaneous pile of leaves, then require the children to sit down and sort them, putting leaves of one kind into one pile, of another
kind into another pile, thus making as many piles
as there are kinds of leaves. Try same plan with
bark.

After an afternoon's outing of this kind there
will be no difficulty in fixing the association of bark
and leaf by reviews in the school room, which may
be done by holding a single leaf, or a single piece
of bark before the school for the judgment of the
pupils.

To fix the habit and habitat of trees, questions

like the following may be asked: Where did you find your walnut tree, Mary? On high ground or low ground? On black soil or red clay? What kinds of trees grow in the same neighborhood with the walnut? Where did you find yours, John? And yours, Susie? Do you find any walnut trees along swamps or marshes? 'What kinds of trees do we find in such places? Where did you find your water beech, Willie? Your cottonwood? Quaking aspen (Quakin' Asp)? Birch? Willow? Are these all neighbors to one another? Which are neighbors of the walnut? I wonder if these neighbors are an advantage or an injury to the walnut. (Here the struggle for existence may be discussed to some extent, taking care to bring out the benefits as well as injuries.) Call attention to the large elm that stands out in some vacant lot, as in a door yard, or open field, and compare with one that grows in the dense woods. What difference in shape of head or top? In the amount of ground covered by its shade? In the density of its foliage? In the general thriftiness of the tree? Why these differences?

Take the walnut also as an example and notice its beautiful top when in the open field. Notice its condition when in the dense forest. 'Account for the difference. Invite the personal experience of each pupil in this respect, letting him name other trees that may have come under his immediate notice.

Wallace says, that in every contest between the birch and the beech, the latter has been most successful. The former loses its branches at the touch of the beech and throws all its strength in the top trying to grow higher, but in this it fails to get as much light and moisture as the beech, two very essential things in the life of a plant; hence the one is driven to the lakes and swamps, while the other holds possession of the field. The beech can flourish in the shade but the birch cannot, hence the advantage of the one over the other. The beech will also kill out the fir for the same reason as that given for the victory over the birch. But the old sturdy oak plants his roots deep in the earth and challenges the beech to mortal combat. The stubborn beech, so long the victor over all other competitors, refuses to yield and withdraw from the field, but hurls defiance at the mighty oak. The contest is on, the battle is long and hotly contested at every point, and at this date, the beech is still undaunted and is maintaining his ground with remarkable tenacity, while the oak shows signs of fatigue and is beginning to weaken. The present indications are that the oak, too, will have to seek grounds where the shade of the beech is still unknown. "The aspen, birch, fir, oak and beech appear to be the steps in the struggle for the survival of the fittest among the forest trees of Denmark." See Wallace's Darwinism, page 22.

When the spring time comes and the buds begin to swell, the teacher should make a collection of sprays from the various kinds of trees studied in the preceding autumn. Attention having already been called in the second year work to the preparation of buds for the winter, a short review of that work will be a suitable introduction to the study of buds at bursting time. To begin, pass several buds of different kinds to each pupil and require the children to carefully remove the outer brown scale in each case. This may be done with a needle or common brass pin.

Notice that all the buds are more or less sticky, but the horse chestnut is especially so; that others are fuzzy, hairy and scaly, and that all these different characteristics are especially adapted for the protection of the delicate green parts of the bud within.

Put fresh twigs of the different trees studied in a jar of water and watch the unfolding of the buds. See which unfolds first, the terminal bud or lateral buds. Which is the larger, the lateral or terminal bud of each of the several trees studied ?

Which has the greater store of nourishment? Why should that be so? What becomes of the scaly covering as the bud unfolds? Watch closely the hickory bud so as to know what is done with the scales. Do the outer coverings of all buds do as the hickory scales do? What change in color of the outer covering in the cherry or apple buds? Where is the flower bud found in cherry? Examine both lateral and terminal buds and note that the latter as a rule do not bear flowers, but leaves. Put twigs in a jar of water and watch the flowers come out from the lateral, but not often from the terminal buds.

Have the children observe the scale scars of last year, and notice that when the scales of this year fall off they will leave the same kind of scar, a sort of scar ring, around the twig or branch. How often then do these scars or rings appear? Answer:—As often as the bud scales fall off, which is usually once a year. Then if a branch has four of these rings, how old is it? Then can we tell the age of a branch? How? There are exceptions to this rule but it will not be necessary to mention them here.

C.—Study Arrangement of Leaves.

After the foregoing study of buds has been quite well completed, attention may be called to the arrangement of buds or leaves on the branch. For this purpose collect very carefully young shoots of normal growth of the several trees studied. Distribute them to the pupils, giving to each as many specimens as kinds of trees. Begin with linden and require each pupil to hold in an erect position, a branch of that kind of tree. Place one hand upon the lowest bud or leaf. Where is the first leaf above that one as related to the lowest? Where the second? The third? The fourth, and so on? Which leaves or buds are vertically above the lowest? Start with any other leaf and see if the same result is true. Try the elm and compare it with the linden as to arrangement. Take the birch, beech, walnut, apple, pear, peach, and note in each case the number of times it will be necessary to follow the spiral of buds or leaves around the stem before reaching the leaf vertically above the first. Now take the maple and box elder and have the pupils state the difference in leaf arrangement between them and the other trees. This will lead to the terms opposite and alternate as applied to the position of leaves upon the stalk. The buds of all maples and box elders are opposite, while those of most other trees are alternate.

At another time collect some gasses or garden plants of any kind and study their leaf arrangement in same way.

REVIEWS:—*Characteristics of Plants.*

1. *Oak.* — Leaves alternate, simple, nett-veined, with stipules deciduous. Flowers generally in catkins. Fruit an acorn. Bark of *red oak* with smooth stripes. Of *white oak*, scaly. *Burr oak*, furrowed. *Black oak*, black bark.

2. *The Birch.*—Flowers in bright, yellow catkins. Leaves of *black birch* heart shaped and doubly serrate. Of *gray birch*, triangular with a long taper point, twice serrate. Of *red birch*, ovate, acute at both ends and doubly serrate. Arrangment alternate. The bark of *black*, or *cherry birch*, with an agreeable smell. Of *gray*, or *white birch*, scaling off in white strips and layers. Of *red birch*, loose, shaggy and reddish brown.

3. *Horse chestnut.*—Leaves palmately compound, and composed of seven leaflets all diverging from the same point on the leaf-stalk, opposite. Flowers are in panicles or racemes, and yellow and reddish in color. Each panicle is as large as a lilac raceme. The fruit is a mahogany colored seed about the size of a hickory nut, enclosed in a prickly burr.

The buckeye is the American horse-chestnut and has five leaflets instead of seven, and the fruit burr is smooth instead of prickly.

4. *The elms* have alternate, simple leaves; straight veined and serrate edged. Flowers in clusters in axils of last year's leaves, purplish or yellowish green. Fruit, dry, winged, or nut-like. Bark fairly smooth in young growth, and very rough and much cracked open, forming ridges of an inch or more in height. *The slippery elm* has reddish colored wood and an inner bark sweetish, mucilaginous and pleasant to the taste. The leaves of the elms vary greatly. Those of the slippery elm are rough on the upper surface and downy on the lower; while those of the English elm, the American or white elm, the corky white elm and winged elm all have leaves smooth, especially above. The *white elm* has abruptly pointed leaves with petioles, while the *winged elm* has small, thick leaves with scarcely any petiole.

The *walnut* includes in its family two genera, viz.: the walnut and hickory. The former has two well-known species in Indiana, viz.: the black walnut and butternut, or white walnut. The whole family is known by its alternate pinnate leaves, no stipules, and sterile flowers in catkins, fertile ones single or two or more in a cluster, bearing single fruits called nuts, enclosed in a shuck, which remains green in color until the fruit begins to ripen, after which time it begins to turn brown, and finally black. The *black walnut* is known by its minutely downy stalks and shoots, and smoothish serrate leaflets. Also by its large, round fruit, from which the dried shuck does not fall away. Bark is very rough and deeply furrowed, even more so than the elm.

The *butternut* is known by its oblong, sticky fruit and sticky leaves. The fruit is about twice as long as thick, and is held within a persistent shuck similar to the black walnut in that respect. The fruit of both black and white walnut is deeply furrowed. Why?

The *hickory* is represented by several species. The *shell-barked hickory* is well known by its shaggy, hard bark, which almost invariably shells from below upward. Its compound leaf has five leaflets, the three upper much larger, and lance-ovate in shape. The nut is white and grows within a shuck, which easily falls away upon ripening. The *large shell-bark* has seven to nine leaflets, which are more downy beneath than those of the common shell-bark. The nuts are yellowish in color and much larger and thicker shelled than the common shell-bark. The fruit of both these shell-barks is edible.

The *pig nut* is known by its rough, not scaly, bark, smooth leaves, leaflets five to seven, obovate, lanceolate, fruit bitterish.

The *sycamore*, or *button wood*, is known by its large leaf, heart-shaped at base, which contains an inverted cup in bottom of leaf-stalk, which covers the newly-forming bud. Its whitish, green bark separates into thin plates, which soon fall away. This tree bears balls which contain seeds.

For further discussion of trees, see Apgar's Trees, Gray's Botany, or Howe's Science Teaching.

4. *Study Growth of Vegetables.—a.* Radish. *b.* Onion. *c.* Lettuce.

Suggestions : Plant a few seeds of each of these plants to be observed during growth. They may be planted in a common soap-box, or, if it be desired to have each kind separate, cigar boxes may be used, though a deeper box would be better for the radish. These should be planted in February, if the school-room is safe against frost, so that the children will be able to get as much of the life history as possible before school closes in June.

During the first year of this course, the children studied the forms and colors of leaves, observed their preparation for winter, their budding forth again in the spring. They also learned to distinguish the parts of a plant, as root, stem, leaf, flower.

In the second year, they studied the forms of leaves, and learned to associate them with the plants or trees on which they grew.

Germination of seeds planted in school-room and preparation of leaves and buds for the winter, effect of frost, etc., were observed, the greater attention being given to the facts easily seen. Now, in the third year, the pupils should be led to see the *function* of things; perhaps not all things, but some things.

They are already learning, under the study of the parts of the flower, that the function of the flower is to make seed. It seems very appropriate to introduce this subject of *Seed Factory*, as given in (2) of this course, after the seeds of radishes, onions, and lettuce have been planted, since the logical inquiry would be, where did these seeds come from, and how were they made?

The seed factory now having been established, as in (2), the inquiry must continue until the function of leaf, stem and root is fairly well known.

How does the root assist in the making of seed? First, it holds the plant in place in the ground. Second, it sucks up moisture for the stem to carry up to the seed factory. In dry times, it is necessary for the roots to dig deeply in order to secure the proper amount of water. The roots of elms have been known to travel a distance of two hundred feet to reach a pond that distance away. Willow roots have penetrated to the depth of fifteen feet, and pushed through board curbing and brick walls to reach the water in open wells.

In 1898, at Harry Pierce's residence on Silver Heights, overlooking New Albany, the four-inch waste-pipe of fire-brick tiling, solidly cemented at joints, became completely obstructed by a solid mass of willow roots, so that the water could no longer find a passage. On examination, it was found that these roots penetrated the cement and entered the pipe through these joints. The tree it-self stood several feet away from the tiling and as much as five feet above. From this, and from many other instances already on record, it is evi-dent that the roots are faithful to their trust as water carriers. They bare their backs, so to speak, and even heave up brick pavements, in their frantic efforts to obey the will of the superintendent of the seed factory.

Let the children narrate instances in their own experience, and bring reports from the parents in regard to the energy of the roots. Call attention to the fact that the large roots provide themselves with tiny root-hairs to suck the moisture out of the very fine particles of dirt; in this way the dirt itself is screened back, thus providing the factory with pure, clean water,—a wonderful filter! Use the hand lens to see the root-hairs on wheat planted on blotting paper, or on cotton in a tumbler of water. These may be seen, however, very well without a hand lens.

Our next attention is to the leaf. Do roots and flowers have to do all the work there is to be done in the seed factory? What are leaves for? If a leaf, the under side of it, should be covered with dirt, what change in color would that part of the plant undergo? Did you ever see the corn-blades and maple leaves covered with dust from the road during dry time in August or September? What effect did the dust have upon the thriftiness of the plants? Did you ever wash the leaves of your house plants? Why did you do so? Could you breathe as well with your nostrils filled with dust? The nostrils, or stomata of plants, are most numerous on the under side of the leaf. Do you now see why the leaves should be kept clean? Does breathing air have anything to do with the growth of our bodies? How? Does breathing air and moisture have anything to do with the growth of the plant? Then does the leaf have anything to do in helping the flower to make seeds?

Call attention to the many thousand mouths (stomata) of the leaf, and the fact that each little mouth is opened and shut at exactly the right time;

also, that the leaves act as stomach and lungs to the plant. What takes place in our lungs? In our stomachs? In the lungs and stomach of the plant?

As there must be a boarding house to feed the men who do the work in basket factory, or iron foundry, so must there be a boarding house to feed the helpers in the seed factory. Now the leaves constitute that boarding house. The sap is brought up by the stem, air and moisture through the mouths of the leaves, and all these elements are worked over by the sun into the choicest food. But these workers cannot leave their respective places to come to the boarding house for their meals, so the food must be sent to them in some way. Did you ever carry dinner to factory men? Well, that is what veins are for, to carry food, and this they do quite cheerfully. Food is in that way sent to the roots, for they have to eat in order to grow and do their work; to the stem, for it, too, has to work in carrying water to the factory and sap to the leaves, hence it must eat; to the seed factory itself, because it also must eat to develop seeds, and the leaves eat what is left after supplying all the others. Refreshments are sent in form of sugar to the seed factory and the seed factory works it over into starch and deposits it in the food part of the seed. Then, when the seed is planted, the sun helps the little plantlet in the seed to change the starch back again to sugar, in which form it readily becomes soluble in water and hence food for the baby plant. What are our baby plants, radish, onion and lettuce that we planted a week or two ago now doing?

To show the influence of the sun in this boarding house, or in the formation of "leaf green," place a plant in a dark place for a few days. What effect upon the color? Upon growth? Do all the helpers in the seed factory get fed as well as when the plant was in the light? How can you tell? What change takes place in the growth of tree or plant after the leaves have fallen in autumn, or in extremely dry weather? When caterpillars have eaten all the lungs and stomach of a plant, what change do you observe in the growth of the plant? What change in growth of seeds? Why? Why should the onion beds, radish beds, and lettuce beds be kept clean from weeds and insects? Why do gardners "thin out" their vegetables? When are they too thick? Point out the leaf, stem and root of the radish, onion and lettuce as you did the other plants. Compare market values of these plants.

B.—ANIMAL LIFE.

1. *Lessons on Snake, Fish, Frog, etc.*—Collect a garter snake, a frog and a fish, keep them alive in

separate glass jars, pass them around the room for
the children to examine. What sort of covering
has the snake? The fish? The frog? Will the
fish and frog shed their skins as will the snake?
The teacher should be able to secure some cast off
snake skins for examination. Do the scales of the
fish assist that animal in its movements through the
water? Do the scales of the snake help in motion?
Which way do the scales slope? Why not forward
instead of backward? Where does a fisherman
begin to scrape the scales from a fish? Why?
Could we scrape the scales from a snake in that
way? Is the skin of the frog scaly? Why would
not scales be an advantage to the frog as well as to
most fishes? Where do fishes live? Frogs?
Snakes? Touch the frog's eye with anything soft.
Has it any lids? Of what kind? Touch the
snake's eye in the same way. The fish's eye.
Why should not these animals as well as frogs and
turtles have lids to draw over their eyes? Try
turtles. Why should not the frog and turtle have
upper lids as well as under lids? Notice the round,
smooth spot just back of the eye in the frog. Do
you find any such spot on the snake or fish? See
if you can find such a spot on the grasshopper.
Look just above where the long leg joins the thorax.
Frogs and grasshoppers can hear, but that matter
is still in doubt in regard to snakes and fishes.

What sort of tongue has the snake? Why does he almost constantly keep darting it out of his mouth? Examine the frog's tongue. Why is it long and sticky? How does it catch its food? Would a fleshy, sticky tongue be of any use to a snake? Explain that the snake does not seek for gnats and flies as a frog does, but for much larger animals, such as grasshoppers, frogs, birds and mice, and that to hold such animals, a sticky tongue would be of no particular use. A frog sits nearly still upon its feet, slightly throwing its body and head forward as it thrusts out its long, sticky tongue to capture a fly or gnat that happens to draw too near. The snake, with mouth wide open, throws the front part of its body forward, and between its wide open jaws, siezes its prey. The lower jaw is composed of two parts, separated longitudinally, each part moving forward and backward freely upon the other. The upper jaw is provided with teeth that point backward toward the throat; also each half of the lower jaw is so provided. Then in swallowing food, the prey, while still alive, is held firmly between the upper jaw and one half of the lower jaw. The other half of the lower jaw is thrust forward for a new hold. The process of swallowing has been compared to a boy in a fixed position pulling a load towards him by means of a rope, hand over hand. In the case of swallowing, it is jaw over jaw instead of hand over hand. In the case of the frog there are no teeth in the lower jaw, neither is the lower jaw divided as in the snake. But the upper jaw is provided with a patch of very minute, short teeth in the roof of the mouth. How does that arrangement assist the sticky tongue in disposing of gnats and flies? How do the mouth and tongue of the fish differ from those of the snake and frog? Why should this be so? How does the snake breathe? The frog? The fish? Why should the fish have gills? Was there ever a time when the frog had gills? Why should not the snake have gills? How many legs has the frog? The snake? The fish? What stands for legs in the fish? The pectoral fins take the place of the fore legs, the ventril fins the hind legs. Describe the movements of each of these animals. Ask the children to watch the toads catch flies, gnats and other small insects under the electric lights, summer evenings. Where is the toad in the daytime? Look for it in flower-beds, gardens, or close along the fence or house, lying flattened out with its nose tucked downward, looking for all the world as if it were dead. How do these animals spend the winter? Hibernation. Are these animals of use to man?

How? Then should they be killed just because they are toads, frogs, turtles and snakes?

Call attention to the fact that there are but two kinds of poisonous snakes in Indiana, viz.:—the rattle snake and copperhead. Should the garter snake be destroyed simply because its brother rattle snake and copperhead do harm? The teacher should try to induce the children to lay aside their prejudices against harmless snakes, frogs, toads and turtles.

2. *The native wild animals common to locality.*—Help the children to make a list of the birds of the neighborhood, distinguishing between those that migrate and those that remain over winter. The following plan may serve to show what can be done in the study of birds:—

ENGLISH SPARROWS.

1. Color of male; of female.
2. Size of male; of female.
3. Where do they live?
4. What do they eat?
5. Where do they spend the winter?
6. Where do they nest?

This study may be made in autumn and verified in spring.

7. Where do they roost at night?
8. Why do they not roost on the ground?

9. Do they hop or run? Why do they not run?

10. Are they of any advantage to man, or do they do more harm than good? How? Teacher, please see Blatchley's Geological Report for 1897, page 935.

Deal with other birds in the same way, until you have collected reports from all the birds in the neighborhood. Compare each bird named with the English sparrow in regard to bill, claws, habits of eating, migrating and nesting, use to man, etc.

THE RABBIT.

1. Color. 2. Size. 3. Where does the rabbit live? When does he come out of his den? Did you ever see him by moonlight? Why is he sometimes called Cotton Tail? What does he live on? When is he fat? Why fat at that time? Why is he very poor in hot weather? Why do we not like to have Cotton Tail in our orchards, among the young trees and shrubbery?

What harm will he do in a garden?

Did you ever see a white rabbit?

How does the white rabbit differ from Cotton Tail?

Let the children tell as much experience as they have about these rabbits. The Kansas Jack-Rabbit may be mentioned in this connection, speaking specially of its long ears, long legs, and enormous size.

In connection with every animal studied, be sure to talk about its manner of gathering food, and use or damage to man.

3. *Study common domestic animals.*—Let the pupil make a list of the domestic animals found in the neighborhood, and compare them with the wild ones already mentioned. Study each animal from the standpoint of use, habit, how cared for, commercial value of animal, value of product, and determine, if you can, where each came from.

C. PHYSIOLOGY.   (Winter Work.)

1.   Review of previous work:—(*a*) Senses.   (*b*) Skin. (*c*) Teeth. (*d*) Nails and hair. (*e*) Bones. (*f*)   Effects of clothing, shoes, etc.   (*g*) Tobacco and alcohol. (*h*)   Breathing.

2.   *Study of foods:*—(*a*) Strength givers. (*b*) Heat producers. (*c*)   Bone builders.

Why does a man need to eat food? Why do children need to eat food?   Is this true of all animal life?

Lead the children to see that every movement that is made is attended by loss to the system.

Let the children apply the fingers to their wrists and count silently their own pulsations in a minute, the time to be given by the teacher. Let each child make his own record on a piece of paper. Also take record of breathing in the same way.

Then require the school to take vigorous exercise of some sort, and take records again.

Why does the circulation increase in rapidity? Why the increase in breathing?

To supply this waste requires food. Foods are of four kinds, viz.: The albuminoids, starch and sugar foods, fats, and mineral foods.

Bread, milk, beans, peas, fish, lean meat and eggs are regarded as strength-producing foods. These are called albuminoids, because they contain albumen, a substance closely resembling the white of an egg.

The starches are found in bread, oatmeal, potatoes, beans, corn and wheat in any form, fruits, rice.

The heat-producing foods are fats from animals, and oils obtained from nuts. By eating nuts, butter on our bread, fat pork or beef, we keep up the heat of the body. In cold countries, like Klondike, it is necessary to eat a great deal of fat. The natives eat the fat of walruses and seals.

What would the natives of Cuba or the Philippines eat? Would they have any use for walruses or whale blubber? Why?

All meats are valuable as food. *Pork* is hardest to digest. It also contains a large amount of fat. Then would it be good food for a hot or cold climate? Why? Would it be good food for a student? A minister? A teacher? A day laborer? Why? Winter or summer food? Why?

Call attention to the fact that some hogs are afflicted with *trichina*, and on that account it is extremely necessary to cook pork thoroughly done.

Salted meats are not so digestible and nutritious as fresh meats. Milk is the best food of all. It contains *casein*, the stuff that cheese is made of, fats, the stuff that butter is made of, and a sugar, known as "sugar of milk." But milk, to be healthful, must come from good healthy cows that are fed upon good food and water. A disease in water or food, when drunk or eaten by the cow, may be transferred to the milk and hence to the boy or girl who drinks it. Typhoid fever has been conveyed in that way to persons using the milk. But pure milk, cream and all, will support life longer than any other food.

Eggs are highly nutritious, but when fried in grease, they are regarded as less digestible.

Wheat flour owes its value to *gluten* contained in it.

In baking bread, yeast is used to cause the bread to rise. Teacher should explain how this is done. Why can we not make as light bread from corn meal as wheat flour? (Small quantity of *gluten*, the tenacious property of wheat.) Corn contains more fat than wheat, also more starch. Which would be the better food for Klondike, corn or wheat? Why?

Beans and peas are good food because they contain so much proteids and starch.

Potatoes contain too much water and not enough proteids to be regarded as the best food.

Radishes, cabbages, turnips and carrots contain too much water and too little starch to be valuable as food.

Has alcohol any properties whereby it may be called a food? Has tea or coffee? What are the claims for these drinks? See Martin's Human Body or any other good physiology for full discussion of these subjects.

What are mineral foods? Why do we need salts of iron and lime and common salt? What kind of food helps to make bone?

3. *Water, why we use it.*—Where may water be found? In lakes, rivers, ocean, wells, dew on grass, rain, fog, etc.

Take a chunk of dirt the size of a base ball and squeeze it between two boards, putting on several pounds pressure, and the boards will become moistened from the water pressed out of the dirt, if the dirt is not too dry.

Take any growing plant, submit it to pressure and the juice will ooze out. How is cider made? Cider, when fresh from the apple, is water holding sugar and other substances in solution.

The juice from the sugar cane, from which we make molasses, sugar and candy, is simply water holding in its hands sugar and candy. What is the juice of the water melon? Cantelope? Peach? Grape? Anything? What is the liquid part of our blood? Where does the liquid come from that we find in a fresh blister from a burn, or from friction caused by hard labor, or a rough place in the shoe rubbing upon the foot? Then can water be pressed out of muscle, skin, fat, or even bone, if the pressure be great enough?

Man is about two-thirds water, "enough," one author says, "if rightly arranged, to drown him."

Food for the plant must be dissolved in water before it can serve as food for the plant.

In our lesson on root hairs we said that these hairs were little mouths* minutely small that reach out to the fine dirt and suck the water out. They serve as a filter, keeping chunks of sand, or lime, or clay, or any solid matter from entering in a solid state into the structure of the plant. But all substances soluble in water can be drawn up

*This is based upon the principle of osmosis, and though mouths do not exist in root-hairs it is thought to be a simple way of presenting the principle to the children.

by these root-hairs and thus get into the structure
of the plant. Dissolve a small quantity of table
salt in a tumbler of water. Can root hairs now
take up salt? Dissolve sugar. Can root-hairs now
take up sugar when they drink the water containing it? Could they have done so before it was
dissolved? Can root-hairs take up grains of sand?
See if sand will dissolve in water. Can root-hairs
take up lime? See if lime is soluble in water.
Then how do lime and sand or silicon get into the
growth of the plant? We know these substances
are there, for they form the strength to the stalk of
wheat, grass, or corn, etc. It has been proved by
experiment that when these substances are found
in proper proportions they unite chemically, thus
forming a substance that will dissolve in water.
Thus united and thus dissolved in water the root-
hairs have an easy task to perform in carrying
these substances up to the factory. What, now,
is the great use of water in plant life? I wonder if
it serves the same purpose in animal life?

Here the teacher should explain that the *villi* of
the intestines are something like root-hairs dipping
into the food which has passed from the stomach,
and drinking the liquid which contains sugar, salt
and other substances in solution. Can anything
not dissolved be sucked up by these very minute

absorbents and carried into the structure of the
muscle or bone?

After the food is sucked up where is it conveyed?
To the blood vessels. What is the liquid part of
the blood? Could new food be carried by these
blood vessels to the lungs and all parts of the system, if there were no water in the vessels? Could
new food be carried into the muscles, be made over
into muscle, to the bones and be made over into
bone, without water? No more than a plasterer
can plaster a house without water. Then what
must be the use of water to our bodies? Lead the
pupils to see that as the great bulk of everything
we eat is water, it will not be necessary to drink so
very much nor so very often. Frequent drinking
is often a mere habit, and ought to be avoided.
Care should be taken not to drink from the shallow
wells in town as they may be more or less contamina-
ted with the privies and other dangerous wastes of
the city. Surface wells and foul cisterns should be
avoided as carefully as you would shun smallpox
or any other deadly poison. Thoroughly filtered
water is the only safe kind, and then it is not safe
if the filter is allowed to become foul. Water
should also be used for bathing purposes. Soft
water is recommended for that purpose. Here,
again, purity should be sought for, because the

skin is porous and therefore an absorbent. If the water contains poisonous gas, or poison in solution, the skin will take it up and carry it into the general circulation of the blood and thus seriously endanger the health of the individual. For further remarks on bathing, see Second Year work and Jenkins' Physiology.

4. *Unwholesome drinks.*—Under this heading all drinks containing Alcohol as an active principle may be considered.

1. Cider. How made? Where does the alcohol of cider come from? What would be the effect of habitual drinking of hard cider? Real hard cider is about one-tenth alcohol.

2. How is alcohol made? Make several experiments with alcohol to show its evil effects. First: put a small quantity in a cup or saucer and set fire to it. Will it burn? Second: treat the white of an egg in a cup or small vessel, with a quantity of alcohol. Stir for a few moments. What is the effect upon the egg? What then would be the effect upon all proteids, or albumenoids in the stomach if treated to a drink of alcohol? What effect upon the brain? Nerves? Muscles? Would a small quantity of alcohol as found in hard cider, if repeated several times a day for a few years, affect our system? How is beer made? What is the

proportion of alcohol found in it? Would it be
regarded as an unwholesome drink? Why? How
is whiskey made? What is the per cent of alcohol
found in it? Is it a wholesome drink? How is
wine made? Does it contain alcohol? How can
you prove it? Place a small quantity of wine in
an evolution flask, to which should be attached a
delivery tube, leading through a cold bath, for
carrying off alcoholic vapor after it has been driven
from the flask by heat very little less than the
temperature of boiling water. The alcoholic vapors
will be condensed as it passes through the cold bath,
and may be received by a beaker or other vessel
placed under the end of the tube. Touch a lighted
match to the liquid gathered in that way and notice
the alcoholic flame; a positive proof that alcohol
exists in the wine. Beer may be tested in the same
way. Likewise cider. What is fermentation?
Give examples of fermentation, such as the changing
of apple juice into hard cider, of grape juice into
wine, of corn juice and its distillation into whiskey,
etc.

All primary fermentations produce alcohol; then,
why does not light bread produce alcohol? By the
action of heat it escapes with the carbon dioxide.

Put a tablespoonful of sorghum and a cake of
baker's yeast into a tumbler of water and set it in

a warm place, temperature of a summer day, twenty-four hours, and watch it "work," fermentation.

The effervescence which you see is caused by the escape of carbon dioxide gas formed therein. The alcohol is formed at same time, but does not escape so readily as the gas unless warmed to a much higher temperature, as in baking light bread.

In secondary fermentation the alcohol escapes, thus leaving acetous acid or vinegar. In the first, or primary fermentation, the sugar in the juice is changed to alcohol; in the second fermentation the alcohol is changed to vinegar.

Alcohol does not exist in the fruit, or sugar, neither does vinegar exist in the alcohol. A fossil made of limestone preserves the form of a once living animal or plant, but the limestone did not exist in the living tissues of the animal or plant; no more does alcohol exist in living corn, or grape, or apple.

Fresh oysters are wholesome food, but fossil oysters would lie rather heavy on the stomach.

Fresh corn is an excellent food, but, when changed by fermentation into whiskey, it is poisonous to the system.

To make this illustration vivid, the teacher should show some fossil plants or animals, oysters, if possible, so that children may see that nature makes a complete change in quality and properties.

The following rules are quoted from Johonnot and Bauton's Lessons in Hygiene, page 49:—

"I. We should not drink cider, because it is the nature of the alcohol in cider to create an appetite for more alcohol.

"II. Cider deadens the senses and tends to make its drinkers ill-tempered and careless about doing right.

"III. We should get our grape juice by eating the healthful and delicious grape.

"IV. When the grape juice has been squeezed out and its sugar turned into alcohol, it is a poisonous drink, and we should not take it.

"V. Home-made wines, if produced by fermentation, are unsafe drinks, because they contain alcohol.

"VI. It is the nature of alcohol, in even the weakest wines, to create an appetite for more alcohol. Thus wine tends to increase intemperance, rather than to decrease it, as some have supposed.

"VII. Fermentation is a part of the process of making bread, but the alcohol is all evaporated out of the unbaked bread. We should not eat bread that is not well baked. It is not digestible.

"VIII. Beers produced by fermentation contain alcohol. We should not drink beer of any kind.

"IX. There is no alcohol in vinegar. We may safely flavor our food with it. Lemons and limes furnish more healthful acids than vinegar.

"XI. Fermentation entirely changes the character of anything it works upon. The germs that cause stewed fruits and preserves to ferment are not the alcoholic ferments, but they spoil the stewed fruit or preserves, and make them unfit for us to eat.

"XII. The habit of chewing tobacco is not only disgusting, but very injurious, as the poison of the tobacco, the nicotine, is dissolved by the saliva and absorbed by the system.

"XIII. The use of cigarettes by young boys weakens the muscles and hinders growth, besides causing other serious injuries. The healthy body does not require tobacco in any form and it should never be used."

5. *Bad effects of tobacco.*—To introduce this subject call for the different forms in which tobacco is manufactured, viz.:—cigars, plug, fine-cut, smoking-tobacco, cigarettes, long twist, etc. Make a lesson or two of each form, showing pupils as far as possible the manner of manufacture. Take a cigar in hand, cut it with a sharp knife down one entire side, thus laying open the filler which in most cases, especially in cheap cigars, is made of scraps, and in many cases, of the filthy sweepings

from the floor of the tobacco house. Collect a few "snipes," old discarded stubs, and let the children smell the sickening odor from the poison that gathers in the mouth end of the cigar during process of smoking. If you can find a very strong pipe, borrow it, take it to the school room and show it to the children; disconnect the stem from the bowl and dig out some of the strong nicotine and let the children smell it. How nauseating it is! How are cigars made, by hand or by machinery? I wonder if the hands of the cigar makers are always clean and free from disease. Plug tobacco may be discussed in somewhat the same way. What is saliva good for? What does the tobacco chewer or smoker do with his saliva? Did you ever hear of "smoker's sore throat?" A great man once died with such a disease. When, do you think, did Gen. Grant learn to use tobacco? Do you think he would advise other boys to do as he did? If he had waited to begin until he was a man, do you think he would have learned to use it? Why? Did you ever know a full grown man to begin the use of tobacco in any form? Besides the evil effects of tobacco upon bone, lungs, stomach, teeth, saliva, brain, and the thinking power, it is well to give some thought to the pocket-book side.

Compare cost of a loaf of bread with cost of a cigar. Compare values to the system. How long would it take a cigar smoker to smoke up a horse and carriage? A house and lot? A suit of clothes? Which would you rather have, a good watch or its value in cigars? The teacher should continue such questions as long as they seem to have the desired effect upon the pupils' minds.

Some discussion of tobacco farming may be of interest. Also, the collecting, drying and shipping.

6. *Hints for health. a. Good eyesight. b. Care of ears. c. Care of throat. d. Necessity for breathing. e. Care of feet.*

LESSONS.

1. *The eye. a.* Make drawing on board showing coats, iris, humors, pupil, lens, retina, and optic nerve. *b.* Let the pupils make drawing, this time copying from the board. *c.* Study eyelids, lashes, and eyebrows as seen about the eyes of other pupils, and determine use in each case. Effect if the eyelashes were cut away; if the eyebrows were shaved off.

Use of the tears. Effect if the tear glands were destroyed and the tears dried up. Effect of dust under the eyelids. Give some directions concerning the removal of dust, cinders, etc., when once lodged under the eyelids. Effect of exposure to light. Long-continued looking at one object. Looking at bright objects like the reflection of light from a mirror, or from white wall or paper. Effect of trying to read or write with insufficient light, e. g.—in dusk of the evening.

Near sightedness and cross-eyes should receive proper attention.

2. *The care of ears.* Next to the eyes, the ears are most important, and should receive our attention. The ear should be taught in its three parts, *external, middle* and *internal.* The outer ear gathers the sound waves and therefore should be clean. The wax is of use in keeping out insects and protecting the internal ear from exposure to cold air. While it should not be removed entirely, yet it should not be allowed to harden at the opening and thus clog up the passage way, or to create too much pressure upon the tympanum.

Children should be cautioned against the practise of screaming or whistling in another's ear. Such practise is dangerous to the ear. Catarrh causes deafness, hence care should be taken to prevent taking cold. Do not pick the ear with a pin or other hard substance. Avoid the use of tobacco, as that will produce inflammation in the ear, and dry the tympanum and thereby cause confusion of sounds.

3. *Care* of the *throat.* Show that the apparatus for the voice is in the throat.

Illustrate by chalk box or other device with fine strings stretched and fastened across the top. Pick the strings between the fingers as you would a banjo. Notice the change in sound when the strings are tightened or loosened. A violin, banjo or guitar will be a still better illustration. In either case the making of a new sound requires change of fingering, tightening or loosening of the string.

What produces the change of sound in our voices? While singing the musical scale let the children place their hands upon their own throats and observe the muscular changes that take place. With a severe cold could these muscular changes be made as easily? If the strings of a violin were covered with a sticky substance would they give us as clear a sound? When the vocal cords are covered with phlegm, as they must be during a severe cold, can you speak as clearly? What tubes belong in the region of the throat? (The tube which leads to the lungs and the one to the stomach, —two very important ones). If the trachea be inflamed, what effect upon our breathing? It produces a cough which only adds to the inflammation, and is otherwise very disagreeable, not only to ourselves but to others. A heavy cold will often cause

sore throat and swelling of the tonsils. Such a
condition injures the voice, hinders easy breathing,
makes swallowing difficult and helps to stop up the
nasal passages. If allowed to continue, chronic
catarrh is unavoidable. The use of tobacco or
stimulants only irritates the inflammation.

Avoid wearing comforters or other warm bandages
around the neck, as such over-care may result in
conjested condition of blood in the throat.

4. *Necessity for bathing.*   See remarks on Second Year work in regard to bathing.

5. *Care of feet.*   Tell stories of Chinese mothers
who bandage the children's feet in order to make
them fashionable.   Make a drawing of the arched
form of the foot, and show that the bones of the
foot must be free to act in a natural way.   What is
the benefit to be derived from the arched form of
the foot?   What effect would a very tight shoe
have upon the arch?   What effect upon the freedom
of the toes?   Then what effect upon easy walking?
Have you ever seen a shoe made so that the little
heel would rest under the hollow of the foot?   Did
you ever wear such a shoe?   Where should the
heel tap be?   Why?   Is the "toothpick" shoe a
good one for the toes?   Why?   Which is more
sensible, a low heel or a high heel?   How high
should the heel be?   How high is your natural
heel as compared with the ball of your foot?

Tight shoes will cause corns, which, when once formed, are very troublesome. Any rough place in the shoe, pressing upon the foot, if long continued, will cause an increase of thickness in the cuticle; such increase, when it presses so hard upon the ends of the nerves beneath as to cause pain, is a corn. A corn may be destroyed by cutting it thin with a knife and removing the pressure.

Tight shoes will also cause in-growing toe nails, bunions, etc., all of which are very painful and interfere very much with the elasticity of step, and gracefulness of movement.

Hosiery with holes in the toes or heels should be avoided, as they are liable to become wrinkled or folded in such a way as to bring on corns. Cleanliness of feet and foot-wear is essential to the health of the entire body as well as the feet.

Rubbers should not be worn in the house, as they prevent the excretions of the feet from escaping. If these poisons do not escape they will be reabsorbed by the skin and thus injure the system.

7. *Ventilation.*—(See breathing in Second Year work.) It is just as necessary for the lungs to be fed as it is for the stomach; and as the stomach requires pure and wholesome food, even so must the lungs have pure air. Air breathed over again

is unfit for healthful respiration. If we live in
rooms so tight that the air cannot be changed, we
cannot avoid breathing again and again what we
have thrown from our lungs as waste matter, and
also the poisonous matter that is constantly pass-
ing through the openings of our skin. How may
we get rid of these poisonous gases and at the
same time receive plenty of pure air instead?

Experiment:—Make two apertures in the same
window by raising the lower sash and pulling
down the upper. Hold a burning taper first at
one opening and then at the other, and note the
result in each case. Close the window and try the
experiment at the upper and lower part of the
open door and note result. Again, close the door
and raise the lower sash of a window high up so as
to give one large opening, try the burning taper
at the upper and lower part of the aperture and
note the result. In each of these experiments let
the pupils describe what the air is doing, and let
them draw the conclusion that the foul air will
leave the room through any opening large enough
to let pure air in. The thing to be avoided is direct draft, or wind, hence care should be taken to
open a door or window on a side opposite from the
wind. To avoid a direct draft in a sleeping apart-
ment, place a board about three or four inches

wide under the entire lower sash; such an arrangement will prevent rain or snow from blowing into the room and will at the same time permit fresh air to enter between the sashes. For sitting rooms, the old fashioned fire place, or open grate, is the best means of ventilation. What makes the fire roar in open fire places? And the flames to wallow up the chimney? (Ascending current of air from the room.) Which is better ventilation, a soft-coal burner with an open hearth or a hard-coal base-burner? Why? Why does the base-burner cause head-ache so often conplained of in warm rooms? (Because the draft through the fire is not strong enough to carry off the bad air of the room and also because carbon monoxide leaks out of the stove and diffuses itself throughout the room, thus rendering the air extremly poisonous and dangerous.)

Enumerate instances of persons poisoned by escaping gas from a base-burner. Tell the story of the Black Hole of Calcutta, where one hundred and forty-six persons were shut up for the night and only twenty-three were found alive at dawn of day. Also, one stormy night, one hundred and fifty persons were crowded into the cabin of a ship and only eighty came out alive. What makes children listless and drowsy in a warm, close school room? Why do you go to sleep at church sometimes? Why do you get up with a headache some mornings? Will it pay to use every effort to get good air in all our rooms?

d. Weather study. Continue work as outlined for second year. See outlines for both first and second years.

FOURTH YEAR.

1. *Rapid review of third year's work.*

2. *Uses of leaves, roots, and sap.*

Leaves. See lessons for third grade, reviewing form, margin, variation, surface, etc., and proceed to the *structure* of the leaf. Very carefully strip off the thin outer covering of any fresh, plump leaf and notice its transparency. Have the pupils do the same with their leaves. Compare with outer skin on fresh, ripe grape; with outer skin of a very ripe apple; with the epidermis of our own skin, and draw the conclusion that the epidermis of the leaf is used for the protection of the inner part. Then the inner portion must be very important to the life of the plant, otherwise it would not need protection.

The epidermis of our own skin, we said in the work for the third grade, has thousands of pores through which perspiration may pass to the surface and escape. Examine the epidermis of the leaf and see whether it is constructed in the same way. It will be necessary to use a good hand lens for this examination. The larger openings called stomata are found on both sides of the leaf, but most numerous below in higher land plants. This discovery leads us to believe more strongly than before that the inner portion of the leaf must

have some important work to do, or else it must
contain in store something of infinite value to the
plant; otherwise it would not go to the trouble of
constructing these stomata through the epidermis, thus making thoroughfares, or gateways, con-
necting the inner green mass with the outside world.
Why does the detached leaf wither when exposed
to dry air or sunlight? Try the experiment to see
that such a result is true. Why do corn leaves
curl up during the heat of the day in dry weather,
and open out again at night? Invert a tumbler or
glass fruit jar over a bunch of fresh leaves, and in
a few minutes observe the moisture collected on
the inside of the vessel. Where did the moisture
come from? Can you now see any use for the
pores of the epidermis? Why do plants look so
bright and green after a warm spring rain? Do
the stomata have anything to do with it? How?

We have already learned that root hairs draw
moisture out of the soil, and with the moisture all
dissolved substances, the water serving chiefly as a
carrier, or vehicle, for conveying material to the aid
of the seed factory.

We also learned that the *villi* of our elementary
canal opened their mouths just wide enough to
drink the dissolved portion of our food, the water
or liquid serving as a carrier for the food that could

not be taken into the system any other way. So it is with the stomata of the leaves. They breathe in order to get carbon from the air. The plant needs carbon to make sugar and starch. In fact carbon enters into the structure of woody fiber, bark and everything about the plant. It is carbon that makes wood such a useful article for heating purposes in our stoves and furnaces. It is carbon that makes beans, corn, wheat, potatoes, etc., useful foods. Thus it is clear that carbon must get into the plant some way, but how? Carbon is a solid and can no more than sand be dissolved in water. But old Mother Nature provides a remedy. She knows that oxygen has easy access to the leaf factory, coming and going continually. She also knows that oxygen is in love with carbon and if permitted to do so, the two will join hands.

In this union carbon is no longer a solid, but a gas, and in this condition effects an easy entrance through the portals of stomata to the plant–substance factory. Old Mother Nature here turns sunshine loose. Sunshine, like a jealous lover, separates the newly wedded pair, and, leading oxygen out through the upper portals in the free open air, leaves carbon a prisoner in the plant–substance factory to work, and form new associates in the fiber, tissues, and seeds of the plant. Oxygen and carbon,

united in the way mentioned above, are together
called carbon dioxide and when dissolved in water it
is called carbonic acid.   It is carbonic acid that escapes from our lungs.   We have no more use for this
acid gas, and hence throw it out with our breath.   The
plants are glad to take it up.   This they do as just
stated, by breathing it through the stomata of their
leaves.   The plant-substance factory is not in
operation at night, so carbonic acid gas is allowed
to escape to the atmosphere until morning.
Throughout the day, if the factory is running at
full speed, carbon dioxide is separated as rapidly as
it arrives, the oxygen returning to the air for more
carbon, securing which, it makes another visit to
the factory; and thus the process goes on throughout the entire plant growth.

The vapor that we collected on the inner surface
of the tumbler in our experiment with the
green leaves a while ago was given off by the
leaves through the stomata.   This process of
giving off vapor is called *transpiration*.   Again
experiment with the plant a few days by putting
some in a dark place, others in a room of medium
light, and others in a strong light for the same
length of time.   Note the difference in color and
in growth.   Account for the difference.   Take
another body of plants and separate it into groups

of equal size, age and strength; from the one group cut off all the blades from the petioles, repeating the operation as often as new blades appear. In every other respect the same favorable conditions should surround both groups of plants. At the end of one week note the difference in growth of main stalk both in thickness and in length. Why do potato beetles injure the growth of the potato? What effect do caterpillars have upon the growth of plums, apples, peaches, gooseberries? Why? The leaf, then, must be a sort of factory auxiliary to the seed factory. It is in the leaf that all the material brought up by the roots or absorbed through the stomata is made over into plant substance. But our experiments prove that this factory cannot work in the absence of sunlight and proper temperature and moisture. Sunlight, then, is one of the factors in this factory. But these leaves do more than take in water. They breathe the same as we do with our lungs.

Leaves also serve as shade for the protection of the delicate young shoots or branches against the scorching rays of the sun. Also to give slope to carry the rainfall and dew to proper places, where the plant can be served to best advantage. In fact the whole upper surface is more or less concave, somewhat like the palm of our hand, and therefore capable of holding water.

Frequently the petiole is grooved upon the upper surface so that the surplus water may be carried to the stalk, or trunk, and hence down the plant, moistening its entire surface. If the rain continues until the leaves are loaded with water, the petioles bend with the weight, the margin and apex of the leaf gradually straighten and turn downward, allowing the water to flow as from a roof, dropping from leaf to leaf, branch to branch, making its way to the very outer edge of the longest limb, from which it pours upon that part of the ground immediately over the finest roots of the tree; thus the leaves, after they themselves are supplied with drink, turn the surplus over to the roots, a very clever act indeed.

SUMMARY:—

Leaves are used (1) for the assimilation of food, making plant-substance; (2) for transpiration of moisture; (3) for respiration, breathing in carbon dioxide and throwing out oxygen; (4) for shade, protecting delicate shoots and fruit from heat of sun; (5) for sloping off the rain so that it may fall within easy reach of the finer roots of the plant; and we may add for shading animals and man from the hot sunshine. At least the shade is so utilized for such purposes. Also for medicine, and tea to drink.

ROOTS.

As an introduction to the subjeet of *root function*, the teacher should collect, or have collected, an assortment of roots, such as radish, to represent the fleshy root, onion to represent fibrous root, the dodder, or mistletoe for parasitic roots. Underground stems such as the potato tuber should be studied to know that they are not roots. Real roots do not bear buds, neither do they have leaves or leaf scars. The eye of the potato is a bud in the axil of a leaf scar, hence the potato is a stem. Other examples, such as Artichoke, the root stalks of garden Iris, Solomon's Seal, Indian Turnip and Trillium should also be examined to show the various modifications of under-ground stems. Bring in a whole "potato hill" if possible, so that the children may see that the roots of the potato plant are very different in structure from the tuber, which is simply a store house for the starch of the plant. Into this store house or potato, the entire life of the plant goes to sleep through the winter, and is called into existence as new plants at the return of spring. The first and most important use of roots is the absorption of water, not for the sake of the good to be derived from water especially, but for the food-salts dissolved in the water. See remarks for third grade.

The second use of the roots is to fix and hold the plant in the ground, or natural sub-stratum. In most cases the same roots perform both offices, as bean roots, corn, ragweed, hogweed, any kind of tree. Corn roots may be an exception, in that after jointing, this plant sends from a lower node a whorl of braces called aerial roots, whose chief office is that of guy rope to the top-heavy corn stalk.

Some plants, such as creeping vines that grow to cover a whole side of a stone or brick church, have two kinds of roots: one kind in the ground to draw nourishment from the soil, the other the clinging roots which attach themselves to the stone wall for support. To prove that these clinging roots do not absorb food supply, cut off the upper part of the vine from its source of supplies in the earth, and that part will soon wither and die.

To prove that the roots in the earth do not give mechanical support to the stalk, remove the vine from its fastenings in the wall, and that plant will immediately fall to the earth.

In fleshy roots, such as radish, turnip, parsnip, etc., aside from mechanical and vital suppport to the plant, a third use is required, viz.: that of storing up starch, fat, sugar and other reserve materials to be drawn upon by the plant at another period in life.

Where is the store of nourishment in cabbage? When does the plant draw upon it to support the seed factory? (The next year.) How about the turnips? The radish? Parsnip? In what respect are these roots like the potato tuber? How different?

See third year work for illustrations of the persistence and energy of subterranean roots, for example, the willow.

Nearly all the fleshy roots are biennial plants, while the annuals are fibrous. This provision is good economy in nature, as the biennial plant must have a rich supply of food on hand to develop the second year stalk that bears the seed, while the annual exhausts all its energy in the development of seed the first year. Where is the store-house of the annual? (In the seed.)

The storehouse of the biennial during the first year is underground. Where is it at the close of the second year? Plant a few grains of corn in a box of dirt. In a few days examine the growth. Which part of the young plant appeared first, the root or the stalk? Why should the root appear first? The first root is called the tap root. When does it cease to be of service to the plant?

Let the children examine the several stalks that still hold the tap root and shell of the grain from

which the stalk grew. What has become of the
food part of the grain? What is now the condition of the tap root?

Can you tell when the tap root ceases to be of
use to the plant? What about its growth after its
use has ceased? A few lessons on survival of the
fittest would be of interest in connection with this
subject of roots.

In a box of radish seed sow also some crow-foot
grass seed, and note the result as to which is mas-
ter of the garden.

On your way to school some autumn morning
pull up by the roots some crow-foot grass, timothy,
Jimson weed, and some other weeds and grasses,
and let the children discuss the merits of each
plant as to its ability in the struggle for existence,
taking into account the adaptability of the roots to
the maintenance of the plant.

THE SAP.

As an introduction to this subject, review stems,
and show clearly what the stem is, as distinguished
from the other parts of the plant? How does the
stem assist the leaves and roots in their work of
nourishing the plant?

Review the purpose, or function of the leaf.
Also of the roots.

Show that the material absorbed in solution by the roots must be conveyed to the leaves for making into plant substance, and that this solution is sap. The leaves receive the sap from the stem, as raw material. By the action of sunlight this raw material and carbon-dioxide are changed into plant-substance and distributed to the growing parts of the plant.

Further than this, sap cannot be discussed in this grade.

3. *Study of useful roots and tubers, as beets, potatoes, etc.—a.* Form. *b.* Structure. *c.* Uses.

This topic has been sufficiently covered in (2) of this course.

4. *Study of grains and grasses, as wheat, corn, oats, timothy, clover, etc.—a.* Harvesting. *b.* Marketing. *c.* Commercial value.

Unfortunately, it is difficult to obtain material of growing wheat, oats and timothy during school year.

If possible to bring these to maturity by planting in boxes, do so, and consider them all at the same time.

Points to be observed:

1. Kind of stem, whether hollow or pithy.

2. Examine the nodes to see if the hollow, or pith, continues throughout the entire length of the stalk.

How do the nodes compare in size to the inter-nodes?

3.   Arrangement of leaves, whether opposite or alternate, and of what rank.   Also, the peculiarity of the bases of the leaves.   Notice that in these particulars these grains are nearly alike, and differing only in size of stalk and manner of producing fruit.

Clover, of course, forms a conspicuous exception to the foregoing, as it belongs to an entirely different family, viz., pulse, and will have to be considered separately.   The others all belong to the grass-family.   How does the termination of the stem of corn differ from that of wheat or oats? Where is the pistil of corn?   Where is the pollen?

Compare each with the others as to root.   Which is the best fitted for battle in life?   Why?

Next take up the seeds themselves and notice that all are covered with a transparent shell for protection.   Note the color of each grain.

Where is the starch located?   Why so much of it?   What part of the grain does the squirrel like best?   Why?   Find the little plantlet of corn. The plantlet of wheat.   Soak the oats and timothy in warm water a few days, and point out the germ in each.   How many ears of corn on a stalk? How many heads of wheat?   Of oats?   Of timothy?

How many rows of corn to the ear? How many grains? How many grains of wheat to the head? Of oats? Of timothy?

Describe manner of harvesting wheat in the first settling of this country. Describe the evolution of the present method of taking care of wheat and oats, including the gathering and threshing.

Method of hay harvesting, stacking, baling. Method of gathering corn in "ye olden times," and at present. Comparative values in the market at home and abroad. What home industries are in existence because of the production of these grains? (Milling, distilling, cattle and hog raising, etc.)

B.—ANIMAL LIFE.

1. *Interdependence of plants and animals.*

Enumerate the animals that depend upon plants for a living, viz.:—Elephant, giraffe, cows, horses, quails, grouse, pheasant, barn yard fowl, etc. What does the elephant eat? cows? horses? quails, grouse or prairie chicken? barn yard fowl? English sparrow? squirrels? rats? mice? rabbits? Such animals are herbivorous, but many of them also seek insects and the flesh of dead animals, but such habit of diet does not excuse their dependence upon plants.

What constitutes the food of cats?. dogs? tigers? panthers? wolves? lion? eagle? hawk? bear? leop-

ard? hyena? Are these animals in any way dependent upon plants? How?

How does the hedge-hog live? How the mole?
the shrew? ground hog? whippoorwill? bats?
These live upon insects. Are they then dependent upon plant life? How? Suppose all vegetation to be destroyed. Describe the scene that
would follow in the animal kingdom.

We have already learned that the animal life
must have oxygen to breathe, in order to relieve
the waste and purify the blood. We also learned
that plant life, under the influence of the sun, is
continually supplying that oxygen. Plant life
demands for its food carbon dioxode; the animal
life furnishes this food to the plant. Thus there is
a constant interchange of food between the two
kingdoms of nature. All the elements necessary
for animals' food is in the soil; then why could not
animals and man live without plants?

Let us illustrate. All the material of our clothing is on the sheep's back or cotton field; then why
do we not go to the sheep's back or cotton field
when we need a suit of clothes?

All the material of our chairs and school furniture, except castings, is found in a sugar, beech, or
walnut tree; then why not go to these trees and
load our wagons with some choice furniture to

take to our homes? What is necessary to be done before the sheep's back will furnish our clothing, or the tree our furniture? Then what must be done before the elements of the ground will nourish our bodies? As the manufacturer and tailor change the wool into clothing, and the cabinet maker the trees into furniture, so do plants imbibe the elements of soil and other foods for animals to eat. How do plants accomplish such a wonderful task?

1. By keeping the original rocks moist, and by decay of leaves, mosses, etc., producing carbonic acid which dissolves the rocks, and thus gives more soil for a more luxurious growth of vegetation.

2. By breaking rocks in pieces by the penetrating power of roots.

3. By holding moisture back along the sides of hills to prevent its running away in floods, until the plant life can have time to utilize the food dissolved in it.

As to manner of food storage, plant substance and seed factories, see topics already written in previous pages of this book. What have animals done to repay the plant kingdom for all this labor? Take the fish-worm for example. The fish-worm, or earth-worm, has so burrowed or channelled the ground that it is difficult to find a place where they have not been.

1. Where ground is tilled, these earth-worms help to carry water from the surface to the tile ditch, and thus help in the drainage.

2. In time of rain after a long dry spell, they help the water to soak into the ground rather than have it run off into ditches, and thereby become wasted so far as the plants are concerned.

3. They reclaim waste land by bringing dirt to the surface and covering up stony places, etc. Henson estimated the amount of dirt brought to the surface to be about 36 tons to the acre per year. Darwin's estimate was about half that amount. On an average, according to Henson, there are 53767 worms to the acre, or 36.5 tons.

The ants, also, do an immense amount of excavating and opening up the soil so as to make it more conducive to the health of the plant. Bees help in the fertilization of flowers and thus pay for all the honey obtained. The same may be said of the hornets, yellowjackets, bumble bees, etc. The butterfly and moth also aid in this, but their larvae are destructive to the life of the plant. But then, again, insect-eating birds destroy the larvae and thereby repay by way of defence to the plant. Birds help in the distribution of plant seed. Animals of various kinds help in the dissemination of seed. How? The excrement of animals and birds make the soil rich and thus nourish the plant. Even their bodies when they die decompose and become food for plants.

SUMMARY.

What plants do for the animals.

1. Plants furnish ready-made oxygen for animals to breathe.

2. Plants gather the elements out of the ground and make them over into food for the animals' subsistence.

3. Plants hold back the moisture on sloping lands and thus prevent arid wastes. By this means gushing springs, to slake the thirst of animals, burst out at the foot of the hills and bluffs.

4. Plants furnish protection from storm and scorching sun.

How the animals repay the plants.

1. Animals yield up carbon dioxide as food for the plant.

2. Animals drain the ground and thus make the soil in proper condition for the plants.

3. Animals (earth-worms) lead the water of heavy rain in dry time through the soil to an underground outlet, rather than permit it to rush away in one general flood and thus do the soil no good.

4. Animals assist in fertilization.

5.   Animals assist in seed dissemination.

6.   Animals give their excrement and dead bodies to the nourishment of soil for the benefit of the plant.

DOMESTIC ANIMALS.

a. Productions of.   *b.* Treatment.   *c.* Uses.   *d.* Products.   *e.* Commercial value.

Let the horse be the type for general discussion, as children are interested in this animal more than in any other.   Under this head may come the different kinds of horses on basis of use, viz.: Draft horse, coach horse, race horse, driving horse, saddle horse; on basis of stock, Norman, Clydesdale, heavy draft, Indian pony, Canadian or Montana pony; Texas pony, etc., with the general characteristics of each.   In connection with the study of the horse, read portions of Black Beauty, by Sewell.

This book gives an exhaustive treatise on treatment. Cows, dogs and other domestic animals may be introduced at many points in the discussion of the horse by way of comparison and contrast. Kind treatment should be the emphatic topic in every case.

The comparative commercial value of cows, fat stock, horses, and so forth, may be discussed, together with cost and care to be deducted in each case.

PHYSIOLOGY.

1. Review foods. See work for third year.

2. *Digestion.*—*a.* Organs of. *b.* Functions of each organ. *c.* Processes. *d.* Favorable and unfavorable conditions.

After a discussion of the foods, as suggested in the Third Grade, put them through a process of cooking, and then introduce the children to the process of *mastication*, the first process in digestion.

Teeth.—Number, kind, action.

Saliva.—Show where it comes from, and what excites it to action. Prove that one of its uses is to enable the jaws, tongue and cheeks to work smoothly with each other.

To do this, have each child bring a clean napkin, and, at the order of the teacher, mop out their own mouths perfectly dry and immediately thereafter try to chew or open the mouth, and work the tongue. What result?

Again mop the mouth dry and immediately thereafter put sugar on the tongue, and see if it can be tasted. Try sand, and see if the child, while the tongue and mouth are absolutely dry, can detect the difference in taste. It must be then that the saliva is used to dissolve solubles in the mouth so as to give the tongue a chance to select the good and reject the bad food.

Again, let the pupil try to swallow when the mouth and throat are perfectly dry. What result? Saliva, then, is necessary to swallowing.

Let the pupils chew grains of wheat for a few minutes. Note the appearance of a sweetish, sugary taste due to the action of the saliva on the starch of the wheat. Note further that starch is insoluble in saliva, and, therefore, has no taste, while sugar is both soluble and sweet.

The saliva, then, is used to change starch into sugar, and hence in solution to make them more pleasant to the taste and to facilitate digestion.

Let the pupils now enumerate the four uses of saliva, and state what effect the use of tobacco would have upon each use.

The food has quite a series of processes to go through before it can be of any service to us. After leaving the mouth, it passes over a little draw-bridge that always falls across the opening to the wind-pipe whenever it sees any solid or liquid coming. You know the windpipe is so constructed that nothing but air, or other substance in the form of gas, can pass along its track without causing violent strangling and coughing. Poisonous gases, at least some kinds, will also cause strangling, and even death, if introduced into the system in any way, but especially through the windpipe.

The windpipe is a tube that leads downward to our lungs, and it is through it that the lungs are filled with air, and through it the bad air escapes from the lungs. So, you see, it is very important that this tube be kept for the passage of air only.

Put your hand on your throat and feel that gristly tube just in front of the gullet. That tube is the windpipe. Now, the gullet you cannot feel with your hand, for it is just behind the windpipe, and extends behind it all the way downward to the stomach. The gullet carries the food, receiving the same as it rolls off the other end of the bridge mentioned awhile ago. Then the bridge flies up again, so that our breathing can continue as before.

The gullet is composed of thousauds of muscular rings around a lining member that is smoother and finer than the lining of your cheek. This inner lining is so smooth and moist that the food, pushed from behind by muscular rings, easily glides along down into the stomach, which is a large room containing capacity for three pints.

How is the lining of the gullet kept moist?

But the greater part of our food is solid, such as potatoes, beans, meat, fruits, etc., and can no more enter into the blood, the general circulation, than sand can be taken into the sap of the plant by the root hairs. So something must be doue. Nature here again provides a remedy. This solid matter must be dissolved before it can soak through the walls of the stomach or be taken up by the *villi*. See third grade work in regard to *villi*. So a set of organs are made to manufacture solvent juices, which are poured upon these solids in the stomach. These juices keep pouring into the stomach until the food is sufficiently softened and liquified to pass out at the back door (phyloric orifice) of the stomach into a long coiled tube called the small intestines. Here again other solvent juices are poured in until all the food that is fit for use in the body is dissolved just like sugar in milk; then the *villi*, like the root hairs in the ground, suck the juice up and the lacteals (milk drinkers), a system of hair-like tubes, carry it onward to a larger tube called the thoracic duct, through which the food fluid is passed into the veins, and thence through the heart to general circulation throughout the body, visiting muscles, nerves and bones, repairing the waste places on the way.

EXPERIMENT.

1. Dissolve a small lump of limestone with hydrochloric acid. Note the rapid action as the limestone passes into solution.

2. Pour a half glass full of water into the acid over the limestone. Note the decrease in action.

Question: When the solvent juices of the stomach are acting upon the solid food just swallowed, what would be the effect of a large drink of water, or any other drink? These solvent juices act best when the temperature of the stomach is about as warm as blood. Then what effect would a glass of ice water or a dish of ice cream have upon digestion if taken just after or during a meal?

Review the organs of digestion together with the function of each, and develop some hygienic laws relative to best digestion and consequent good health. A lesson or two on the effect of alcohol upon the stomach and its juices, would be altogether proper with this subject of digestion.

A general discussion of "stomach troubles" such as dyspepsia, may be deferred to the eighth grade.

3. *Circulation.* *a.* Organs of. *b.* Functions of each organ. *c.* Favorable and unfavorable conditions.

We stated a while ago that the dissolved food passed through the thoracic duct to veins, thence to the heart, and from that organ to the general circulation. Just at this point a good beef heart from a slaughter house should be provided and the teacher should proceed to dissect the same in the presence of the pupils, showing the right and left auricles, right and left ventricles, valves, entrance of veins and departure of arteries, distinguishing between those of the pulmonary system and those of the systemic. The heart may be represented as a pump, forever receiving blood from the lungs and body, and continually pumping it again to various parts of the system. (See third year work.)

Color of blood? When a finger is cut, what is the great natural process that stops the flow of the blood? (Clotting.) What is the use of blood to the system? How does it create heat? How does it promote growth and repair waste?

"Blood, then, is a very wonderful fluid.' * * * But you will not wonder at it when you come to see that the blood is the great circulating market of the body, in which all the things that are wanted by all parts, by the muscles, by the brain, by the skin, by the lungs, liver and kidneys, are bought and sold. What the muscle wants it buys from the blood; and so with every other organ and part, as long as life lasts, this buying and selling is forever going on, and this is why the blood is forever on the move, sweeping restlessly from place to place, bringing to each part the things it wants, and carrying away those with which it has done. When the blood ceases to move, the market is blocked, the buying and selling cease, and all the organs die, starved for the lack of things which they want, choked by the abundance of things for which they no longer have any need."—*Foster.*

d. Weather Study—See outline.

FIFTH YEAR WORK.

A.—Plant Life.

1. *Study stems as to arrangement and growth.*—
Collect a number of stems from trees of different
kinds, selecting the younger shoots as far as possible. Each pupil should be provided with several
specimens. Point out the bark, wood and pith.
Where is the pith? The bark? The wood fiber?
These parts will be better fixed in mind by drawing
cross-sections of the stem. Require the pupils to
draw cross-sections of walnut, hickory, oak, elder,
heavenwood and some of the herbs for comparison.
See if the children can separate the bark into three
parts, viz., outer layer, middle, and inner or bast
layer. Compare several sprouts, one of which
shall be the linden or basswood. It will be interesting now to give a general classification of stems,
so that the pupil may know them in all their forms,
bearing in mind that the stem is the axis of a plant,
which, when developed, always bears geometrically
arranged leaves, and that a branch is a secondary
stem. For the purpose of such classification, collect some plants, as sage, lilac, elder, hazelnut,
haws, crabs, or thorns, raspberry, rose, pea vines,
beans, grapes, pumpkin vine, strawberry, little
dew-berry, passion flower, Virginia creeper, ivy
(avoiding the poison ivy, which is known by its

three leaflets), morning glory, hop, rushes and the tall grasses, tendrils, spines or thorns.

Also underground stems, such as tuber, or potato, scaly roots of various kinds, and many others if the teacher can continue the interest along this sort of work.

Another classification, perhaps more simple, may be mentioned, viz.: Hypocotyls,* scale leaf stems, foliage stems and floral stems. Hypocotyls include all stems whose bud of the main shoot is developed from the apex, for example, the maple, or almost any forest tree, bean, etc. Scale leaf stems are those which are beset with scale leaves and are almost always under ground with axis vertical, as lilies, tulips, hyacinths, onions, stars of Bethlehem. The tuber, a modified scale leaf stem, as also are rootstocks and creeping stems. The foliage stem is above ground and bears leaves with green blades. The peculiar style of the plant is dependent upon the foliage stem. Foliage stems are herbaceous, woody, nodose, scapers, or flower stalks. Erect foliage stems, as caudex, culm, stalk and trunk. The floral stem is that on which the flowers are borne, and may be the main or lateral axis. All stems have nodes, commonly called joints, and in-

*Hypocotyl is the same as caulicle. Caulicle is the first internode, or portion of the stem below the cotyledons and above the radicle or beginning of the true root. It is seldom applied to the part after the plant has developed.

ternodes, the part between the nodes. In all kinds
of stems presented, the nodes and internodes should
be pointed out.

Arrangement of leaves and branches.—The arrangement and mode of branching and form of the
tree must depend, to a great extent, upon the arrangement of leaves upon the branch. As the leaves
are arranged, so are the branches, because the branch
always appears from the axil of the leaf. Opposite
leaves produce opposite twigs and branches. The
maple is an excellent example of such an arrangement. In alternate arrangement, the branch has
but one leaf to each node. Not more than one
branch or twig will appear as a rule in the alternate
arrangement. Compare the linden with the maple,
as to arrangement of leaves and branches. Compare with corn, wheat and other grasses. How
many vertical ranks in each specimen compared?
Try to find stem with three ranked arrangement.
Try cherry, poplar, apple, peach, plum, and count
the ranks (five ranks.) Try the yard plantain for
eight ranks. How about the potato? How do you
know that it is a stem and not a root? Have roots
nodes and internodes? Have they ranks in budding lateral roots? What is a stem? What is a
secondary stem? Observe that it is the business
of stems to produce roots, leaves and fruits.

    2. *Study buds as to position, arrangement,*

growth and purpose. Collect, as in the other grades, a great variety of twigs, especially of hickory, lilac, Balm of Gilead, cottonwood and sycamore. To avoid an over abundance of brush and litter, request the pupils to cut the twigs only three or four inches long and bind them into bundles of a dozen twigs each before bringing them into the school room. As to position, buds are either terminal or lateral. The lateral buds are borne upon the sides of the branch and are usually found in the axil of the leaf. There are superposed buds and adventitious buds on some plants, but it will be as well to avoid the necessity of studying either in this grade.

The arrangement of buds is dependent upon the arrangement of leaves. Compare terminal buds with axillary buds, and account for the greater growth from the terminal bud.

Cut cross-sections of terminal buds and require the children to draw cross-sections greatly enlarged, so as to show the arrangement of the leaves and scales.

Cut cross-sections of lateral buds and draw. Draw stems containing leaves with buds in axils, and also terminal buds showing relative sizes. Draw sycamore bud with the protecting leaf just pulled off at one side, with cup in base of petiole in full view.

In the spring time notice that the lateral buds of the cherry contain the flowers, while the terminals do not, as a rule.

Notice in the dissection of these several kinds of buds that some are sticky and gummy as in the Balm of Gilead or cottonwood, and that those of the hickory and horse-chesnut are woolly or velvety.

Why should this be so? Compare with other buds of out-door life. With buds of green-house life.

In this way, the teacher can easily bring out distinction between scaly buds and naked buds, and that the purpose of scaly buds must be for protection against cold, wet and insects.

Do all axils bear buds? Why? Examine several before deciding.

Buds sometimes lie dormant for a year or more. What will bring them out? In case of accessory buds, which is developed? (The central and larger one.) In the event that one dies, what will be done by the others? What appearance will that give the tree? Do the buds (eyes) of the potato come from the axils? Examine the potato very closely for rudimentary leaf scars.

What are the little black heads found in the axils of the tiger-lily? What are onion sets? When do buds prepare for winter? What signs do trees give us of approach of winter? (Full development and coloration of buds, the coloration of leaves whether frost comes or not.) What have the leaves been doing all summer? (Making food for the plant.) What color is maintained during the factory season? Green from chlorophyll. This now becomes disorganized, and the food is carried away by the sap to be stored up in buds for next year's growth. This disarrangement, or breaking up of the chlorophyll, is the cause of the variety of color in the leaves.

When winter is drawing nigh, the buds hurry into winter quarters. To get ready for this change, the buds make demand upon the plant for all the supplies it can possibly furnish. It demands what is already in the leaf and all the young bark can spare.

The sap begins immediately to make the transfer. In fact, it has been doing a little of that work all summer, for we know that the buds started to build as soon as growing weather set in. But now, along in September and October, the demand from the buds has increased to such an extent that the sap cannot furnish the leaf laboratory any more material, so that the factory must "shut down" and permit the food already made to be carried away to build up the bud.

Coloration of leaves now begins, and as soon as

the sap is done visiting the leaf or when it brings away the last load, the gate to the foot-stalk is forever closed and the leaf is ready to fall. If frost should come and find the gate open a raw place would be left where the leaf joined the branch, out of which sap would flow, and thus lose part of the winter's food; but the plant seems to understand the situation, hence the closing of the gate or sealing over the place for protection. Why do leaves fall after a heavy frost? Are leaves of any benefit after they have fallen? How? What is the form of the food as stored in the bud for the winter? Starch. What change is brought about in the starch when the bud begins to swell in spring time? Sugar. Why is it necessary to change form? So that the sap can dissolve and use it. See lesson on saliva in fourth grade.

The purpose of the bud is to furnish winter quarters for plant or flowers or both. The buds on the tree are a community of stem germs (baby plants).

Compare cocoons of caterpillars with the hibernation of Johnny Jump Up or Spring Beauty. Where is the store of food in each case? Where in the tuber? Compare with hibernation of animals, such as frogs, snakes, earth worms, etc., with the migration of birds.

All nature acts upon one common plan, varying

only in accordance with the peculiarity of the life dealt with.

3. *Study in proper season, exogens, endogens, ovary, anthers, pollen, germination, and the use of the pod.* Make collections of exogen cuttings not more than three or four inches long, tied in bunches as suggested for the collection of buds. In fact the same collection will answer. What is the arrangement of woody matter in each case? Exogens have a bark made of an outer protective layer, a middle layer, and a green bast, or inner layer. Let pupils find these layers from a cross section. In the spring of the year when the sap is beginning to work, the cells out of which bark and fibre are both made, are so tender and soft, being filled with sap, that a very little pressure on the bark will cause the same to slip from the woody fibre. Boys sometimes make willow-bark whistles in the spring time, because of that fact. The growth of the stems comes from that layer of cells, one side building new green bark, the inner side building a new woody fibre. Plants that increase in size in that way are called exogens.

Still examining cross sections, let the pupils observe that within the bark just described is woody fibre composed of bundles arranged in rings around a central pith. See if all this is true of the apple branch, of cherry, peach, sycamore, bean. Have pupils count the rings of growth in stumps of trees and cross sawed timber. Do these rings represent periods of growth? Can we then form some notion of the age of the tree? By way of contrast, make a collection and study of endogens. As in exogens collect bundles of cuttings three or four inches long. Examine cross sections to see if the bark is the same as that which covered the exogens. Can you in any way pound or rub the bark of a cornstalk loose as you did the willow bark? Can you the grass bark? The wheat bark? Sugar cane? The palm? Pond weed? Cat tail? Indian turnip? Banana? Lily? Pineapple? Any of the rushes? Fish pole? Do you find in any of these cuttings the fibre bundles arranged in rings around a central pith? Compare with oak, box-elder or bean in this respect. Make drawings of cross sections of endogens, alternated with exogens so as to get the distinction fully fixed on the mind. Plant a few acorns, beans, peas, beech nuts in same dirt with corn, sugar cane, wheat, and notice difference in number of seed leaves as they begin to come through the dirt. Those having two seed leaves are called dicotyledonous plants, and those having but one seed leaf monocotyledonous. To which of these does corn belong? Apple? Is the walnut exogenous or endogenous? How can you tell?

Oats? Why? Then are all exogens dicotyledonous and all endogens monocotyledonous plants?
Make a list of each kind. Which do you think is
the more useful to man? Why?

Ovary—See lessons on parts of the flower as
given in third grade. Collect a supply of flowers
for the pupils, being particular to select those containing large ovaries or seed pods, so that the children may see with as little difficulty as possible.
Flowers of indeterminate inflorescence will be best
for the reason that some of the lower ovaries are
ripening into quite large pods, while the upper portions are just bursting into flowers. This will show
the progress made by the seed factories. Review
the parts of the flower as in third grade. This may
be done with any simple flower for two or three
exercises before dealing minutely with the organs
of reproduction.

Now as to the ovary. What part of the flower
does the pistil occupy when single? What is their
position when there are two pistils? When there
are several? What part of the pistil does the ovary
occupy? What are the other parts of the pistil?
Purpose of each? Open the ovary of a fresh flower.
What does it contain? (Ovules.) Open the ovary
of a well matured seed pod. What does it contain?
(Seeds.) How do seeds differ from ovules? What
caused the ovules to grow into seeds? The life, or

principle of the seed was introduced by the pollen, afterward the sap carried food from the roots and leaves and fed this new life. This new life is the superintendent of the seed factory. What part of the flower bears the pollen? How does it get out of the anthers? When must it get out in order to give this life to the ovules? (When the stigma is moist and sticky, ready to receive the pollen.) Suppose the pollen waits until the stigma dries off, then what will happen to the ovules? (They will shrivel up and the flower will die.) Suppose the pollen to fall out before the stigma is ready to receive it, what then will the ovules do? The pollen then must neither be tardy nor too hasty if it expects to give the ovules life.

Sometimes insects carry pollen from some other flowers to give life to ovules of a flower, whose pollen has been too late or too soon. In that case ovules do not shrivel, but grow into mature seed. Teacher should here explain the process of fertilization by bees, butterflies, humming birds, etc., allowing the children to assist in the discussion as far as possible. Kindle the interest of pupils into daily observation of insect-life among the flowers.

Collect pods from shepherd's purse, mustard of any form, bean, Judas tree, locust, pea, monkshood, maple, Jamestown weed, or any others, and develop the use of the pod. Collect some real green

ones, to show the condition of the seed when it begins to grow. What are the dangers to which the
seed is subject? What is the use of the hull on
the hickory nut? The pod on the chestnut?
What is the condition of the seeds when they are
ready to burst from the pod? See hickory nut,
chestnut, hazel nut, beech nut, pea, bean, locust
seed, and the seeds from any other pods. What
then can be the use of the pod?

4. *Purpose of Plant.* *a.* Reproductive, to produce seed. *b.* Commercial; use to man.

Let the pupils make a list of vegetables and
plants that have commercial value.—The commercial value is always based upon use to man. The
turnip, cabbage, potato, radish, and other fleshy
vegetables and roots, will be mentioned. Why did
these plants produce such fleshy vegetables? The
turnip next year will produce seed. In the production of seed, the great mass of starchy material
is absorbed, thus leaving in place of the turnip,
a tough leathery shell. How was the turnip
produced this year? What will the turnip produce
next year? What then is the real fruit of the
turnip plant? What must be the purpose of the
plant? To reproduce itself by means of its seed.
What we know as turnip is a great store house
from which to draw supplies for the seed factory
next year. What we know as cabbage is simply a

store house from which the plant next year may
draw supplies for the devolopment of cabbage
seeds, which when sown will produce new cabbage
plants.   The potato, in its native country, is used
in the same way, but man and animals have taken
advantage of these plants, broken open their
store houses, robbed them of their supplies, and
made merchandise of them.   But man is not a rob-
ber after all, for he repays the plant for all that he
has taken.

How does he do this?   How do other animals
repay the plant?   In dissemination, etc.   See interdependence of plants and animals, fourth grade
work.   Fine fruits, as apples, cherries, strawberries,
peach, plum, are used by the plant for the same
purpose of reproduction.   Their flesh is edible and
colorations attractive, both of which lead to dissemination of seed for the plant as well as commerce
for man and food for animals.   Draw the conclusion that self-preservation is the first great law of
nature, and that in obedience to this law, the plant
produces seed and fruit in order that it may live
again; also, that when it yields up its fruit and
store of nourishment to man for his food, the purpose of the plant is not defeated as it may at first
appear, but strengthened, in that man now becomes
the friend of the plant in its struggle for existence.

B.—PHYSIOLOGY.

• (For winter study, at least when not seasonable for plant study.)

1. *Bones.*—*a.* Names and groups. *b.* Forms and economy of structure. *c.* Uses. *d.* Observation of real bones.

2. *Joints.*—*a.* Kinds and movements. *b.* Ligaments and attachments. *c.* Uses. *d.* Observation of action of joints.

3. *Muscles.*—*a.* Forms and structure. *b.* Tendons and attachments. *c.* Uses. *d.* Observation of action of muscles. *e.* Exercise and its relation to muscular development.

4. *Respiration.*—*a.* Organs ; name, form, structure and purpose of each. *b.* Processes; inhalation and exhalation. *c.* Purpose of respiration. *d.* Observation of action of respiratory organs. *e.* Ventilation and breathing.

See any School Physiology on these subjects. The outline is deemed sufficiently elaborate for any teacher.

C.—WEATHER STUDY.

Weather, record of preceding grade continued, noting thermometer, barometer, winds, clouds, rainfall, snow, hail, dew, and frost. (See outline for fourth year.)

SIXTH GRADE.

A.—PLANT LIFE.

1. *Provision of nature for matured seeds.*

(*a.*) Nature provides a store of nourishment for each seed so as to give the new plant a start in life.

The seed-leaves, or cotyledons, as in bean, pear, pumpkin, etc., are packed with food, which occupies the entire space inside the coats of the seeds. Examine these seeds after they have been soaked in luke-warm water over night.

The cotyledons will easily separate, revealing the *plumule* and *radicle*. The cotyledons, plumule and radicle are together called the embryo. All the food there is in such seeds as those just mentioned is crowded into the embryo. Let the pupils mention as many seeds as they can think of that have the food stored in the embryo. ' Next, let the children observe other seeds—corn, for example, after they have been soaked sufficiently long to begin to germinate. The children will notice that the food is around the embryo, like the albumen of the egg around the yolk. Hence the name *albuminous** seeds, that is, seeds whose embryo is surrounded by food supply. The bean, pea, corn,

*The word *endosperm* would be a more scientific term.

etc., are called *exalbuminous*, because they have no food supply outside of the embryo.

To which class of seed does wheat belong? Why? Maple? Hazlenut? Buckwheat? Peach? Apple? Beechnut? It will be a good exercise to let the children germinate a great many seeds and classify them.

(*b.*) Nature provides protection for the matured seeds. As soaking over night in water will facilitate the study of protective coverings,—prepare beans, corn, peas and other seeds in that way, and begin the observation. Pull off the coverings and try to find an outer, hard covering and an inner mucilaginous membrane. Compare the soaked coverings with those that have not been soaked. Is the covering a good protection against insects? Against changes of temperature? Against moisture? Scrape off the covering of five dry beans. Place these five naked beans, with as many covered ones, in a pan of water. In a few hours examine them, and note difference in the result of the experiment.

How do you account for the difference? While the outer coat of the seed is almost impervious to water, still, if no water could enter the embryo, no growth could take place. So nature provides one very small entrance to the inner portion of the seed. See if the children can find it.

It is through this orifice that the ovule was fertilized by the pollen, and through this orifice that moisture penetrates the seed to dissolve the sugar which the sun has made from the starch already in store. If it were not for this opening the bean, or grain of corn, would lie in water many days before showing signs of germination. The outer seed coat has a great variety of colors and forms, ranging from black down through the shades of gray, brown and sometimes yellow, white or red.

The wing-like and hairy appendages, curved, pointed and barbed processes are but modifications of the outer seed coat. What is the purpose of the parachute of the dandelion? Of milk-weed? These are but modifications of the outer seed coat, or *testa*.

Do these parachutes serve any purpose other than distribution by the wind? They serve the purpose of guiding the seed point end downward so that it can work itself into the soil. Let the children observe this fact. Also iron weed, goldenrod, etc., may be observed and discussed in the same way. Notice the long, delicate silky hairs on the seeds of willows. How do they serve the seeds in dissemination? Do they serve any other purpose? Notice the muddy places which are adapted for the growth of willows catch these seeds while the dry places, not adapted for willow growth, are swept bare of these seeds by the wind. The silky hairs are peculiarly adapted for clinging to the mud, spreading out and thus bringing the seeds into close contact with the mud for germination. Notice the maple seed with its butterfly wings, the linden with its skiff, the roundness of walnuts for rolling down hill, the roughness and wartiness of some seeds to hold the dirt well, the mucilaginous coating of the sun flower seeds and gourd seeds, when fresh, to enable them to adhere well to the soil.

2. Show how seeds separate from plants.

a. Fruits that do not open at maturity, fall by the force of gravitation, when the sap ceases to bring food from the leaves, but they fall mos rapidly during windy weather. Why is that true and of what advantage to the seed? Name as many fruits as are known to separate in this way. In this collection, apples, peaches, plums, pears, cherries, berries of every description, of course will be mentioned, together with some discussion as to the wisdom of nature in providing the juicy pulp and colored skin; but the fruits equipped with hairs, curved bristles or hooked spines, together with the winged fruits and plumes should have a prominent place in the discussion.

b. Fruits that open and discharge their seeds when ripe.

They do this in a variety of ways.   The spring beauty bursts open at the top, splitting downward on three sides, so that the seeds can be blown out by the first gust of wind, leaving the dry capsule clinging to the stalk.   The blood-root pod is also erect and splitting from the top in two places, ready to throw its seeds as the stem nods in the wind. The poppy, a twin brother to the blood-root, opens a dozen or more slits with a suddenness that causes its ten-thousand very fine seeds to jump out of the capsule into the wide, wide world, each seed to begin a new life.   The shepherd's purse also splits downward from the top.   Why do so many fruit capsules open from the top?

What kind of weather is best suited for such openings?   Did you ever know capsules to burst open in wet weather?   Why do they not?   How do the seeds get out of the capsules?   Other fruits and seeds may be studied as far as time and interest will permit.

3.   Show how seeds are disseminated.

Many suggestions have already been made under (1) and (2) but perhaps there are teachers who would like the work given in more careful detail. With this thought in view the task is attempted.

DISSEMINATION.

Call attention to the great amount of wild lettuce

found now-a-days along public highways and waste places. How did it get there? Collect some wild lettuce and examine the implements for travel through the air. After breaking prairie sod, especially in the low ground, cottonwood shoots spring up quite plentifully, even when there are no trees of that kind within a radius of several miles. How do you account for it? Examine the provision that nature has made for floating the seeds and find your answer in that.

The first jimson weeds in America sprang up with the tobacco around Jamestown, early in the 17th century. Where did they come from? How? To-day they are found in many rubbish piles and hog lots all over America. How do you account for such a widespread distribution? Here, again, an examination of the burr should be given. What other seeds are distributed in like manner? The pupils will name cockle-burr, Spanish needle, beggar lice, many achenia such as buttercup, hepatica, anemone, etc.

Prof. D. S. Kelly, of Jeffersonville, Indiana, furnishes the following interesting account of weed distribution in the west:

" Not until a few years ago was the *Solanum rostratum* found east of Colorado; but when travel began eastward along the old trails through Kansas, this weed began to work its way along these trails toward the Mississippi.

"The fruit, a berry, is covered by a very prickly calyx, and easily adheres to the tails of horses and cattle very much like cockle-burrs. Now the weed is found abundantly along the main roads as far east as central Missouri.

"In general the migration of our weeds has been westward, but the *Solanum rostratum* is a notable exception to the rule.

"D. L. K."

What seeds are distributed by the wind? Pupils will name dandelion, iron-weed, goldenrod, milk-weed, thistle and others. What seeds are distributed by water? Nuts of various kinds. Cocoanuts, for example, have been carried from island to island in that way; also algae and off shoots from weeds of lake and ocean. Many seeds are carried down rivers and creeks at time of great freshets and distributed along the low submerged lands. Another way of distribution consists in the manner of separation from plants, as described in (2) of this year's course.

4. Show how seeds escape from the ovary. See (2) of this year's work.

5. Show how seeds are protected through the winter.

Berries, nuts, grapes and all indehiscent fruits

are usually left alone by nature, covered with leaves. Ask the children their experience in winter, if they have traveled through the woods. If they have no experience to relate, it will be well to send them through the woods in dead of winter, or better towards spring to find beech nuts, acorns, hickory nuts, walnuts, hazel nuts, crabapples, wild plums, wild grapes, etc. It is more than likely that fully half of the number gathered in this way will have already lost their vitality for germinating, but of those which are alive, nearly all have been nicely covered with leaves as a protection from frost and water. Then, again, the seeds are more or less oily and thus offer resistence to cold and wet. In the majority of cases the outside hull or shuck keeps off the water and sudden frost.

But there is another kind of fruit that does not fall with the coming of frost. See (2) of this report.

Many of this kind of fruit open by means of valves, teeth or pores, and frequently the seeds are persistent long after the splitting of the pods. And as these pods generally open at the apex, provision must be made to protect the seeds against rain or snow.

This is done by a peculiar quality of the pod, which attracts the water in such a way as to form a film of oily moisture over the seeds, and this film, if left undisturbed, will prevent the water from

flowing in and destroying them. It also serves as a
protective covering against frost. When thoroughly dried off again, the first gust of wind sends the
seeds whirling into space, some falling into good
ground, where they are covered by drift, leaves,
etc., there to remain until the sun reaches high in
the heavens in springtime. Then what happens?[1]
But most of the seeds are picked up by the birds,
or otherwise destroyed. Observe an ear of corn
with husks remaining. Corn has been known to
stand out all winter and endure extremely cold
weather, and yet at planting time in spring, be
chosen for the farmer's seed. How is the ear protected? How does the ear usually hang? Would
a farmer choose an ear that stands erect upon the
stalk? Why? Why is corn that has stood all
winter in the shock not good seed?

Notice the covering of seeds. See (1) of this
course. Notice the oily nature of most seeds; take
flax for an example. Unless seeds become soaked
with water, frost will have but little effect upon the
germ. Notice that cockle-burrs, thornapple, ragweed, and almost all other weeds keep a supply
of seed upon the stalk all winter, dropping
them in early spring when dry, windy weather
sets in.

Of what advantage is that to the plant? How
many have noticed sycamore balls persistent

through the winter? Can the seed of the sycamore be better protected on the tree? How?

Protection against animals should here be discussed. Illustrations of these may be found in all
fruits covered with thorns, prickles, and spines.

Examples:—Thornapple, or Jimpson burr, cockle
burr, chestnut, sand burr, beggar lice.

2. North American pine bears cones, the scales
of which are pointed with sharp spines, which de-
fend the seeds against the attacks of animals, until
the seed ripens and falls.

3. Wild roses retain their fruit upon the bushes
long after maturity, when they are distributed by
black birds and other birds that prey upon the
pulp. Mice love the fruit very much and, if permitted to do so, would eat the heart out of every
seed. But the rose bush has thorns or prickles all
along the bark so as to prevent any of the rodents
from climbing after the seed. If the rose bush
should drop its seeds early, what dangers would be-
fall them?

4. The sun-flower seeds are protected from in-
truders by prickly bristles covering the stalk from
bottom to top. How about black berries? In all
these cases which way do the prickles point? If
upward they seem to protect against browsing
animals; if downward, against climbing animals
and caterpillars. How are pods, as bean or pea,

protected from mice? (Long pendant stalks).

After cherries have fallen upon the ground the earwigs and caterpillars devour them. How are they protected against the attacks of these animals when growing upon the tree? (Suspended upon stalks.) Fruits without spines or protective bristles, as apples, cherries, peaches, etc., when unripe, are bitter and therefore not sought by animals. When ripe these fruits are much sought by birds and other animals which carry them away, thus helping in the dispersal of the seeds.

6. Study how embryo gets out of coat. Plant seeds of different kinds such as corn, beans, peas, acorns, flax, etc., in moist sand or saw-dust, in flower pots, cigar-box or glass tumblers, being careful to cover each vessel with a glass slide or piece of oil paper to prevent evaporation. Thick blotting paper, moistened and placed on a pane of glass, furnishes an excellent ground work for the planting of seeds. Sheet cotton would also answer the purpose to a very great advantage. After the seeds have been planted the same should be covered with another sheet of moist blotting paper and then set away in a warm place of the temperature required for growing seeds. The growth of these seeds may be observed at every stage. Let the children see where the coat in each case bursts. In the case of the corn, which end of the plantlet bursts out of the coat first?

Why should that be true? Which end bursts out first in the bean? In the pea? In the acorn? Beechnut?

7. What provisions the plant has for developing. Review the store of nourishment in each of the seeds planted, distinguishing between exalbuminous and albuminous seeds. See discussion of this in a previous chapter. What are the two halves of the bean used for? Of the pea? The acorn? Why do not the acorn and pea rise above the surface of the ground and form the seed leaves as the bean does? How about pumpkin seed, squash, watermelon and corn? Wheat? Where is the food supply in each case? What does the young plant do for food after the store of nourishment is all used up?

Have pupils point out the parts of the young plantlet even in the dry seed before it is planted, and then notice day by day the development of seed leaves, plumule, caulicle and roots, and finally the wasting away of the store house of nourishment.

8. Study roots, stems, in relation to flowers as organs for taking in, etc. See lessons already given in preceding grades.

9. Study forms of roots, leaves and stems. See third and fourth grade work.

10. Examine and compare leaves, seeds, prickles, etc. See work for preceding grades.

11. Study measurements of leaves, tendrils, etc.
Have a sensitive plant, *Mimosa pudica*, if possible,
brought in the school-room, and let the children
touch the leaflets and observe them to immediately
close together, and after a time regain their original
position. Why does the plant behave in that way?
This explanation of leaf movement, either from
touch or influence of light or darkness, rain or sunshine, is too difficult for a work of this kind. For
full explanation of plant movement, the teacher is
referred to pages 532–539, vol. I, of Kerner and
Oliver's Natural History of Plants; also vol. II,
Goodale's Physiological Botany, pages 397–424.
See another chapter of this book for sleep movement of plants. Darwin says that all growing parts
of a plant are in constant movement around one
common center from right to left, or the opposite
direction, from left to right. The tendril and
twiner are only exaggerations of this common
movement. Encourage the children to observe
direction of grape tendril, pea climber, bean, wistaria, hop vine, morning glory or any other vine in
the vicinity, and report.

Can a plant bore its roots into the ground any
easier by this worming process than by straight
pushing? Try to push a tender sprout or tender
twig straight into the ground. Now try to work it
back and forth. Which is the most successful way?

Then which would be best for the root? The movement of the root is so very slow that we cannot see
it move. We cannot see a plant grow; but we know
that it does grow, for we can measure it at different
times and prove it to be true. So, too, we may
measure the movement of a plant around a common
axis. Erect a stationary wire or wooden frame around
a growing plant, say a corn-stalk, being careful not
to let the frame touch the plant in any way. Let
the frame rest upon the ground and support a ring
two or three inches in diameter, encircling the apex
of the plant. Measure from time to time the distance of plant from the ring and it will be found to
vary. At one time nearer the north, then east,
south and west, until it comes around to the north
again. These measurements and records, if correct,
will prove the circumnutation of growth. Try
other plants in same way.

THE FERN.

Upon what sort of soil do ferns grow? Do you
always plant them in the shade? Why?

Examine the blades, or fronds, and learn to describe them. Draw several fronds, being particular
to represent all their parts. Where is the stem
out of which these blades grow? The stem is an
underground rootstock or rhizome, somewhat like
iris, Solomon's seal, calamous or rhubarb. Notice

a young frond just coming out of the ground, watch it until it is fully matured and describe its peculiar action.

These ferns do not produce flowers as all the other plants you have studied, consequently they have no seed factory; but they have another way of reproducing themselves, which is very interesting as well as mysterious. Look on the under side of your fern frond. All ferns are not alike in this respect, but in all probability if you look closely you will find generally along the mid-rib and larger side-veins what at first appears to be a kind of papillae. These papillae, or usually conspicuous bodies on the under side of the pinnae, are called the *sori* (singular, sorus) or fruit-dots. Observing their position, notice the thin scale-like covering, *indusium*. But if you remove the indusium you will find attached generally by very delicate stalks, an oval or spherical body. These free ends are spore cases which contain a powdery mass known as spores. This mass of spores is very similar in appearance to the powdery cloud that escapes from a ripened "puff ball." To the naked eye, these spores have no form, but by the aid of the microscope, it is clear that they are just as definite in form as this round earth upon which we live. Strange to say, too, each little spore contains the possibilities of a new plant, just as seeds do in the flowering plant, but not exactly in the same way. There are many of these spores in one single spore-sac and the spore-sacs are so numerous and small that you ean scarcely count the number upon a single frond. When the sac bursts the air is always dry and considerable breeze stirring. Under such conditions, what is done with the spores? These spores, of course, take the direction of the wind and are sometimes carried great distances, falling upon the ground, here and there, all along the way.

If all ground were good for the growth of ferns, the spores from a single plant would soon populate the entire globe with ferns; but all ground is not good for ferns. These plants are a little particular about the sort of ground they grow in. The soil must be damp and shady—not too wet, but damp or moist. Around old, rotten logs or stumps these plants will grow very luxuriantly, if the soil is damp and shady. These spores, then, when they find lodgment in these damp places, gradually develop into very small discs or flattened leaves which spread out upon the damp ground. Each little flat leaf, or *prothallium*, as the botanist calls it, sends into the ground many very fine and delicate root-hairs. As soon as the prothallium becomes thoroughly fixed by its root-hairs, so as to obtain nourishment for itself and the young plantlet which it is about to give rise to, it develops on the under-surface,

i. e., in monoecious prothallia, two organs called
the *antheridium* and *archegonium;* the former
answering to the stamens or anthers in the flowering plant, and the latter to the pistil or ovary.
In other words, the *antheridium* is the *male* reproductive organ of the plant, and the *archegonium* the female reproductive organ. As the
pollen-tube of the flowering plant must reach
the ovule for fertilization before the latter is capable of development, even so must the energizing influence of the spermatozoid fertilize the egg
cell of the archegonium. Soon after this union, if
the weather is favorable, a young fern plant starts
out from the archegonium, and as soon as the fern
is firmly established the prothallium withers away.

The bean plantlet, you remember, draws its first
nourishment from the cotyledons or seed leaves,
and does not depend upon the soil until the
best part of the food in these leaves is used up.
The same may be said of beechnut or any flowering plant. In the young corn plant the *albumen gives it the first start. Now the prothallium
of the fern answers that purpose precisely. Two
growing points start out from the archegonium at
the same time; one is the leaf point and the other
the root point; but the leaf point grows the faster,
obtaining its food from the prothallium. As soon as

*The word *endosperm* is preferable.

the root point has developed roots and leaves sufficient for an independent existence of the fern
plant, the prothallium withers away just as the
seed leaves of a bean do under similar conditions.
It will be seen, then, from the foregoing description that the fern plant does not reproduce a fern
plant immediately, as do beans, peas and other
flowering plants; but that it first produces spores,
the spores produce prothallia, and the prothallium
in monoecious specimens gives rise to male and
female organs, and finally these organs reproduce
the fern plant, which again produces spores. Such
a process is called *alternation of generations*.

MOSSES.

Where do they grow? On which side of tree do
you find them most abundant? On which side of
rock? On which side of hill? If found on the
.. south side of rock, tree or hill, what must necessar-
ily be the other conditions?

While examining the rocks and trees you will
doubtless find a scaly-like plant that spreads out
over the rock and sticks very closely to it. This
plant is known as one species of *lichen*. Some
people call it moss, but it is not true moss. You
will find it quite plentifully distributed over the
surface of old marble slabs in grave yards. After
many, many years, perhaps centuries, the lichen

causes the rock to decay and crumble into dust or soil,
sufficient to give the moss-spores lodgment. So moss
takes the place of lichens just as soon as the latter has
prepared the soil for the support of the moss forest.
But moss will grow upon damp soil, wherever shaded
and supplied with proper amount of heat. In fact,
it does not require a great deal of heat or moisture
to support some kinds of moss, as it is well known
that great quantities of moss grow in the cold re-
gions of the north where few other plants can exist.
It will also do well upon dry knolls. Let the children state where they have seen moss, and name
some things peculiar to its nature. The Pigeon-
Wheat moss is perhaps the commonest kind and
can be found in great abundance by the children
upon almost any side hill and often between the
bricks of walks where the place is damp and shaded.
The Pigeon-Wheat moss is known by its long
pedicel bearing on its apex a capsule containing
spores which, when ripe, are thrown out to be
scattered by the wind the same as fern spores.

These spores are no more the nature of seeds
than fern spores are seeds, or puff-ball dust is seed.
All seeds contain an embryo and food supply, the
latter being either within or without the embryo;
but the spores contain no embryo. They give
rise to green web-like threads, called protonema,
which have no roots, but cling very closely

to the ground, or brick, or whatever their home may be, until finally, after branching profusely, they give rise to very tiny lateral buds, which, in time, grow into moss-plants that bear the reproductive organs called *antheridia* and *archegonia*. The fern plant is asexual, while the moss plant is sexual.

Though the fern plant has neither male nor female reproductive organs, it is at the same time able to produce spores which fall to the ground and develop *sexual* plants called *prothalia*, from which sexless or asexual fern plants once more arise.

Now the history of the moss plant, though produced by "alternation of generations," is somewhat different from that of the fern. In the moss, the more conspicious moss plant is sexual. It bears male and female organs, and an egg-cell is fertilized by a male element. The fertilized egg-cell remains attached to the mother plant and develops into a tiny sexual stalk which bears on its apex the special reproductive cells or spores. These spores fall to the ground as did the fern spores, and there grow into a usually thread-like structure, the *protenema*, from which the sexual moss-plants arise from buds.

In the fern asexual generation was the more conspicuous; in the mosses, the sexual generations are more conspicuous.

Let the pupils examine mosses and ferns so as to compare and contrast the parts of the one with the other, taking first, the stem; second, the root; third, the leaf and fruit.

SEVENTH YEAR.

A.—PLANT LIFE.

1. Review and continue the study as outlined for the sixth year.

2. Study flowers whose floral envelopes are more or less grown together, etc. (*a.*) Review parts of the flower as in third grade work, and call attention to the essential organs and their relation to the perianth. It must be borne in mind that the essential organs are stamens and pistils, the former bearing the male elements, the latter the female elements ; that the business of the flowering plant is to produce seed, and that fertilization is necessary to the production of seed.

WHAT IS FERTILIZATION ?

Let the pupils be supplied with an ample supply of flowers containing conspicuous stamens and pistils. As far as practicable, the flowers should be of the same kind, so that, when you give directions, there may be nothing to hinder any of the pupils from careful observation. Lead the pupils to see that the stamen is made up of a filament and anther, and that the anther is the pollen pod just as a pea pod is a *seed* pod.

With needles open the pollen case and observe that it is filled with a powdery mass called pollen, each grain of which is just as definite in form as a grain of corn is definite, or a mustard seed is definite in form. As the form of seeds varies in different plants, just so does the form of pollen vary in different plants. So certainly is this true that the plants may sometimes be determined by the shape of the pollen alone.

For examples of these different forms, see the lily with its smooth, oval pollen grains; the sunflower with its spherical pollen grains, beset with prickly projections; the musk plant pollen with its spiral grooves, the evening primrose with its three lobes as large as the central body, and so on, the pollen grains of each kind of plant differing in form from those of all others. But these forms cannot be seen without the aid of a good microscope. The other parts of the flower can be seen very well with a good hand lens. Every pupil ought to have such a lens in order to get best results from an examination of these delicate parts of the plant.

Note the opening of the anther. Does it open on the inside next the pistil, or on the outside away from the pistil?

Now examine the pistil. Review the parts of pistil as in third grade work and note the relation of stigma and pollen to each other and to the ovules within the ovary.

Of what advantage is the sticky substance on the stigma? When is such substance most conspicuous?

It will be a good experiment, if you have a microscope, to place some pollen grains in warm, sweetened water under a cover glass and watch the growth of the pollen tube. The pollen grain, when it lodges in the sticky substance upon the stigma, begins just such a growth, sending its tube down through the style of the pistil to an ovule in the ovary. Through an orifice in the ovule this growth continues until its energizing influence is felt in the *embryo-sac*, where seed growth begins by cell division.

Bergen says, "The process of fertilization is the union of the essential contents of two cells to form a new one, from which the future plant is to spring." The one cell is formed by the elongation of the pollen tube, the other is the oösphere in the embryo-sac of the ovule. No growth of seed can take place without fertilization.

During the growth of the pollen tube, how did it obtain its food supply? From what source? Where did the food come from in the experiment under the cover glass? Could growth of pollen cell have been produced down the tube of a dry style? Could growth have taken place under a dry cover glass? Now explain fully, as far as

studied, all the conditions favorable for the perfect
union of male element with female element, or fertilization.

In Indian corn what is the style called? Where
are the stamens and what are they called? Why
do not full ears of corn grow upon isolated stalks?
Examine a few ears from solitary stalks and see if
the required number of grains is developed. From
your observation what conclusion can be formed as
to the provision of the corn plant for self-fertilization? How are the styles, "or silks," of the corn
protected from the reception of its own pollen?
Examine a stalk of corn so as to be sure that one
or more of the blades pass between the tassel and
the silk, thus acting as a roof to ward off the pollen.
These blades may also serve to keep off too much
water from the stigma, or too much hot sunshine,
either of which would destroy the function of the
stigma, but it keeps off the pollen as well. How
are the chances of fertilization improved by community plant life? How is the ragweed benefited
by community life? Let the pupils examine the
ragweed and locate the seed pod and also the pollen.
Be sure to observe that the pollen is dry and light,
and therefore easily carried by the winds.

What other plants bear pollen that is light and
dry and easily carried by the wind? We may call
such plants *wind-fertilized plants.*

In the spring time collect spring beauty, mustard of any species, hepatica, buttercup, marsh marigold, wind flower and dandelion, or almost any spring flower, and examine the stamens as to whether the anthers are borne upon long, slender filaments, or are they without filaments, i. e., anthers sessile? Does the anther rest upon the tip of the filament, or does the filament adhere to the anther along its entire length? Notice particularly the position of anther in relation to stigma. What sort of opening has the anther in each case? Discuss the advantages and disadvantages of each opening and each position in relation to fertilization.

Now notice the stickiness and heaviness of the pollen as compared with the light and dry pollen of corn and ragweed in autumn. What conclusions do you reach as to method of fertilization amongst these spring flowers? It must be remembered that some flowers do not open until fertilization has taken place. For example, many species of violet bear closed inconspicuous flowers, and these produce the greater part of the seed. Self-fertilization in such cases is the only method. See Darwin's *Cross and Self-Fertilization in the Vegetable Kingdom;* also Geddes & Thomson's *Evolution of Sex.*

Besides the flowers just mentioned, most of

which are polypetalous and regular in form, there
are myriads of others that are gamopetalous and
irregular; also many that are irregular and
polypetalous. For instance the irregular conspicuous blossoms of the violet may serve as an irregular polypetalous flower. Have the pupils
supplied with a sufficient quantity for class examination. Notice first of all the arrangement of the
reproductive organs. See that the five short
stamens connive around the style. Is the position
of the stigma one that would be conducive to self-
fertilization? Does the protruding stigma turn in
the right direction to receive the pollen to best
advantage? Examine the pollen to see whether
it is sufficiently light and dry to be blown from
flower to adjacent flowers by the wind? Then
is it liable to be fertilized by the wind? Is it a
flower that is often visited by bees, or other insects?
What inducement does it offer for the visitation of
bees? Look in the base of the petals for nectar
cups. Look into the spur. Does the beard in the
corolla assist the bee in any way while it is reaching after the nectar? Watch the bees at their
work among the flowers and report what you observe. As the bee passes from violet to violet
could it carry the pollen from the one to the stigma of another? How? Here it will be well for
the teacher to turn aside from the study of flowers

and give a few lessons on the bee and its apparatus peculiarly adapted for carrying pollen. Have the children examine with a good hand lens the third leg of the bee. The first tarsal joint bears regular rows of stiff straight hairs, on which the pollen grains are collected. The stickier and heavier the pollen the better will it adhere to the bee. Comstock says of the bee: "The worker is our common acquaintance, the dull-black and gold-colored companion of our walks, that we watch with interest as she ransacks the flowers of a garden or a wayside for her booty of nectar or pollen, now bending low a violet or clover blossom, now plunging headforemost into a hollyhock or lily, from which she emerges dusty with the gold of pollen doors which barred her way to nectar chambers."— *Comstock's Manual for the Study of Insects, page 674.*

Returning to the flowers, is there anything about the shape and arrangement of the violet flower that injures its chances for self-fertilization ? Is the pollen suitable for wind-fertilization ? Why ? Do these hindrances to self-fertilization and wind-fertilization become helpful to insect-fertilization ? How ? Study bleeding heart, columbine, larkspur, lily of the valley, snap dragon, horse-chestnut, lilac, white clover, red clover, milk weed, pea, bean, locust and wistaria in same way. The

pupils will notice that all these flowers are more or
less irregular both as to arrangement of stamens
and pistils and as to floral envelopes. They can
also be led to see that whenever an irregularity injures the chances to self-fertilization it at the same
time facilitates insect-fertilization.

The author is not quite sure that the statement just
mentioned is always true, but it surely is true in
many cases. For instance, the showy strap-shaped
corolla of the sunflower advertises the nectar in the
slender tubes of the flowers within, and thus the insects are attracted in great numbers. The lower
lip of a gamopetalans corolla serves as a resting place
for the bee, while it gathers the nectar within. The
stamens of the horse-chestnut protrude in such a
way as to form a perch for the bumble-bee while
he sinks his suction pump down into the nectar,
and while doing so, does not pollen cling to his legs,
to be deposited upon the stigma of another flower
which he is soon to visit? It will be observed that
in such flowers the stigma and anthers of the same
flower are seldom ready for fertilization at the same
time. The stigma of one flower visited by the bee
is withered, and therefore in a non-receptive condition, while the pollen, ready to perform its part in
the process of fertilization, clings to the legs of the
bee, and is thereby transferred to a receptive stigma
in another flower, the stamens of which are not yet

ready for action. The stigma and pollen must ripen at the same time, else self-fertilization can not occur. See Gray's Structural Botany, pages 216–242.

In regard to the coalescence of the parts of the floral envelope, it will be well to require seventh grade pupils to learn the botanical names. When the whorl of petals is more or less united, the corolla is called *gamopetalous*, a name signifying grown together. When not united, or when the petals are separate to the very base, the corolla is called *polypetalous*, a name signifying many separate petals.

When the outer whorl of the floral envelope is united throughout any part of the sepals, above the point of insertion, the flower is called *gamosepalous*. When the sepals are separate to the very point of insertion the flower is called *polysepalous*.

The expanded portion of a petal, or that which answers to the blade of a leaf, is called *lamina* or blade; that which answers to the petiole of a leaf, or filament of a stamen, is called *claw* or *unguis*. When this claw is absent, we may say that the petal is sessile.

The lower part of a gamopetalous corolla is called a *tube*, if the sides are nearly parallel or at least not too spreading. For example, the lower part of the morning glory blossom is a *tube*.

The upper part of a gamopetalous corolla, whether divided or united, is called the *border*, or *limb*; for example, the flaring part of the morning glory corolla, or the divided upper part of phlox, is a *border* or *limb*.

A gamopetalous corolla is said to be *salver-form*, when the limb or border is abruptly spreading upon a long, slender tube; as in phlox. It is said to be *tubular*, when the border does not spread, as in the corolla of the trumpet honey-suckle.

The corolla is said to be *wheel-shaped* when the border upon a very short tube spreads out like a wheel, as in bitter sweet and potato.

The *funnel form* corolla, as the name implies, is shaped like a funnel, as in morning glory or bindweed, where the tube gradually enlarges upward from a narrow base and expands outward into a wide border.

The teacher should collect a series of flowers, ranging all the way from a slight coalescence of parts to a full united corolla, and proceed to distinguish the different botanical names by the study of the different forms under observation. Deal with each flower something after the following method:

Description of calyx, whether polysepalous or gamosepalous. If gamosepalous, is it tubular, notched or cleft. How many cleft? Color. Description of corolla.

Polypetalous or gamopetalous? If gamopetalous, what is the form of the corollo? Is its form favorable to self-fertilization? Explain fully. Notice its color as to whether it would be attractive to insects. Notice its odor and nectar, guides and cups, and state whether you think these would assist in insect fertilization? Does the corolla serve as a protection to the reproductive organs? How?

Describe the stamens, especially in their relation to the stigma, which must also be described. Notice whether the stamens and pistils are equally fresh and vigorous at same moment of time. If so, what conclusion as to their adaptability for self-fertilization? Do the stamens look fresh and vital and the stigma limp and withered? If so, what inference as to the ability for self-fertilization? Are the stamens withered while the stigma yet remains fresh and erect? Give inference.

Is the pollen of a kind that will admit of wind-fertilization? Now review all the parts about the flower that offer any inducements to insects, and state all the points favorable to insect-fertilization.

3. Study clustered flowers leading to the compositæ. This study includes the whole topic of *Inflorescence*, or mode of flower arrangement.

Flowers are said to be *indeterminate* when they are situated in the axils of leaves, as in shepherd's purse; and *determinate* when they are from termi-

nal buds.   Either of these forms may be single, or
solitary, or grouped in clusters.   When in clusters,
the indeterminate take the form of *raceme*, as in
the lily of the valley, where the flowers, each
upon a separate foot-stalk, are loosely scattered
along the floral axis; corymb, as in hawthorn or
trumpet creeper, which is but a slight modification
of the raceme, the separate flower stalks being
lengthened in such a way as to allow the flowers
to rest in the same horizontal plane.

What change would have to be made in the lily
of the valley in order to assume the form of a
corymb?   Imagine a currant flower-cluster to be
changed into the form of a corymb.   Explain the
process.·

An *Umbel* is a flower cluster whose floral axis is
wanting.   The flower-stalks or pedicels of each of
the several flowers arising from the same point on
the recepticle carry the flowers to the same height
as in the corymb.   The parsnip and cherry are
good examples of an umbel.   Imagine a cherry
flower-cluster in the form of a corymb; what change
took place?· Imagine a raceme of shepherd's purse
in the form of an umbel; what change occurred?

The plantain flower cluster consists of a lengthened axis along which sessile flowers are closely
set.   Such a flower-cluster is called a *Spike*.   How
must the plantain be changed in order that it may

take the form of a raceme ? Of a corymb ? Of an umbel ?

Shorten the axis of the plantain until the sessile flowers arise from about the same point and the flower cluster would be called a *Head*. The button-wood or sycamore ball is a *head*. How is a *head* different from an *umbel?* A *spike* different from a *raceme?* The *spadix*, as in Indian turnip, the *catkin*, as in birch, willow or alder are simply different modifications of the *spike*.

Broaden the receptacle of the head and surround it with a set of bracts called involucre and you have a composite flower. Examine the dandelion, iron weed, ox-eyed daisy, dog fennel and sun flower. In what respect are these flowers different from a clover blossom, or button-wood ball?

The teacher may be able to trace the determinate inflorescence to the compositae in the same way as with the indeterminate.

The *cyme* of the determinate corresponds with the raceme of the indeterminate. In the determinate inflorescence the flowering is centrifugal, that is, the oldest flower is in the center and the order of flowering is outward. In the indeterminate inflorescence the mode of flowering is centripetal, that is, beginning on the outside and flowering toward the center.

The *Fascicle* is a cyme with flowers much crowded, as sweet williams.

A *Glomerule* corresponds to the *head*. The flowers in a glomerule expand from the center outward, *i. e.* centrifugal, while the flowers of a *head* expand from the outside toward the center, or centripetally.

Does the sunflower expand centripetally or centrifugally? How about the clover? The iron weed? The lily of the valley? What steps would you imagine a sunflower to take in order to become a spike? Which form of flower cluster is most conducive to self-fertilization? For wind-fertilization? For insect-fertilization? Why?

Give an opportunity for extended discussion upon the values of each form relative to fertilization. Let the discussion be based upon field observation of insects at work as well as upon adaptation of form.

4. Become familiar with several of our common families of flowering plants. All plants that bear flowers are called flowering plants. We have families of people, families of animals, and strange to say, families of plants. We group together all plants that resemble one another in flower, seed and fruit, and say that they belong to a certain family. For instance, we say that the bean, the red bud or judas tree, the honey locust and the

wild indigo all belong to the same family, the *Pulse Family.* Can you tell why? Collect these flowers and see if they are alike. See if you can find any other flowers that belong to this family. What are the points of likeness? When these fruits ripen collect the pods and notice the similarity in pods and seeds. Of course these pods can not be collected until later in the year. Would you place the clover in the same family as the bean? Try the pea, wistaria, sweet pea.

Study the common mustard and note that it has four petals, four sepals and six stamens—four long and two short. Taste its stem and leaves. Get a collection of radish flowers, and notice points of likeness between them and the mustard. If possible get turnip and cabbage flowers. Let pupils bring in any other flowers having the same points.

These flowers all belong to the *Mustard family*, sometimes called the *Cress family.* There are over six thousand species or members of this family known to botanists. We cannot expect to make the acquaintance of the entire family, yet it is possible to learn a few characteristics or marks that will enable us to know one of them when we have the opportunity. If the children will go out on the hillside they may be able to find the toothwort, which may be known as mustard by the common marks, and as toothwort by its root and leaves.

Examine kale and report on the marks and taste.

Send two or three pupils down into the bottoms and swampy places after varieties of cress.

Make drawings of each variety found.

Review points common to all members of the Cress family; to all the members of the Pulse family.

Name the points in each species that distinguish the plant under examination from all others of the same family.

If you have the time study the *Rose family*. This is a large family and is divided into several tribes or divisions. The child will be astonished to learn that the apple and strawberry are members of the same family, but they are.

The plum belongs to the same family with the hot-house rose. Can the children tell why? So do the cherry and blackberry. Help the children to find the points in common. Let them search for other members of this family.

Let a pupil describe a plant that has been studied. From the description let the other children identify the plant and name the family to which it belongs.

Again, name a flower and let the pupils describe and name the family to which it belongs. Vary the exercises as the interest and attention seem to require.

Learn the points or marks by which any member of the *Lily family* is known. Examine the tulip, the trillium and dog-toothed violet. What are the points in common? Let pupils collect others of the same family. Is the calla a lily? The Indian turnip? Examine them in comparison with each other and in contrast with the lily.

If these families are well worked out the pupils will be fairly well equipped for the future study of flowers.

B.—PHYSIOLOGY.

1. Study somewhat in detail, foods, hunger, thirst, cooking, fatigue, etc.

2. Study the special senses, seeing, hearing, etc. See third grade work for discussion of these two topics.

NOTE.—Use charts to give correct ideas of the organs of the special senses. Make strong the hygienic teaching in connection with the various senses and topics considered.

EIGHTH GRADE.

PHYSIOLOGY.

If the work outlined in Physiology in the lower grades has been properly done the pupil will have, when he enters the eighth grade, a fairly good knowledge of the structure of the body, of the various processes of digestion, circulation, etc., and will understand the more common demands and laws of health. In the eighth grade the authorized text is to be in the hands of the pupils, and the work is to receive attention and energy equal with that given to other substantial branches of the course.

The work will, of course, treat in the main of subject matter gone over in the earlier grades, but the treatment will be more comprehensive and technical than has been possible up to this point. The consideration of individual structure and function will now give way to the consideration of relation and interdependence. At every place possible the pupil will be led to see the adaptation of structure to function and of both to hygienic demands and laws. As far as possible the laboratory method should be used. Actual material should be brought before the class whenever it can be used to advantage in explaining the structure or function of different organs or tissues. Microscopic study should be pursued whenever it can be done to advantage and will aid to clearer comprehension of physiological facts. Physics and chemistry, in so far as a knowledge of them is necessary to make clear the processes of Physiology, should be studied in connection with this subject.

The following outline is suggestive for the work:

I. *Study the Framework of the Body—The Skeleton.*

    (*a*) Composed of bones and ligaments. Articulation of bones, the synovial fluid and its use.

    (*b*) Structure, chemical composition and use of bones as levers.

    (*c*) Relation of their structure and properties to their uses. Growth and repair of bones, effects of food, exercises, habits, etc. Care of the bones.

II. *The Muscles.*

    (*a*) Structure and properties of muscular tissue, the voluntary and involuntary muscles and their arrangement. Nervous control of muscular action.

    (*b*) Changes that the muscle cells undergo during contraction, carbon dioxide, uric acid and other waste material, the oxidation of food substances, heat.

  (c) Special adaptation of structure and properties of muscles to the ends they
serve; use many illustrations.

  (d) The care of the muscles. Exercise and
health. Effects of alcohol and narcotics.

III. *The Digestive System.*

  (a) Foods, kinds of, and the composition
and value of each. Necessity for food.
Alochol and narcotics not foods. The
value of water in digestion.

  (b) Structure and function of the teeth, stomach, intestines and other organs of digestion.

  (c) Chemical changes in insalivation, chym-
ification and chylification.

  (d) The lacteals, structure and function.
The portal circulation, special features
and importance; assimilation of food.
Adaptation of structure and function.

  (e) Hygiene of the digestive organs. Se-
lection of food, etc. Effects of alcohol
and narcotics.

IV. *The Circulatory System.*

  (a) The blood, its composition and use, properties of, structure and use of corpus-
cles. Coagulation and its use.

(*b*) The lymph, its composition and use, circulation of lymph. Arrangement and structure of lymphatic vessels.

(*c*) The circulation of the blood, how carried on, the heart as a force pump. Action of the valves of the heart.

(*d*) Structure and function of the heart arteries, veins and capillaries. Relation to each other, capillary attraction, etc. Changes the blood undergoes during circulation.

(*e*) Hygiene of the circulatory system. Effects of alcohol and narcotics.

V. *Respiratory System.*

(*a*) Composition of air, diffusion of gases, evaporation, heat, animal heat, radiation, etc.

(*b*) The skin as a respiratory organ. Structure and function.

(*c*) Inhalation and exhalation, how carried on. Organs of. Structure and function of respiratory organs. Adaptation of structure to function.

(*d*) Relation of respiration to circulation and digestion.

(*e*) Hygiene of the circulatory organs. Ventilation and heating, exposure, dress, etc.

VI.   *The Skin and Kidneys.*

  (*a*) Structure and use of each. The skin as
         an excretory organ. Modifications of
         the skin. The kidney as an organ of
         excretion. General view and neces-
         sity of the kidney.

VII. *The Nervous System.*

  (*a*) General character of nervous stimuli,
         need of a coordinating process in the
         body.

  (*b*) Structure and function of brain, spinal
         cord, nerves, ganglia, etc. Action of
         each part. Divisions of the brain with
         special reference to structure and func-
         tion.

  (*c*) Reflex action, its general character. Re-
         lation of nervous system to other pro-
         cess of the body.

  (*d*) Hygiene of the nervous system, with
         special reference to sleep, exercise,
         morality, alcohol and narcotics.

  (*e*) Relation of nervous system to the mind.

VIII. *The Voice. How Produced.*

  (*a*) Structure and function of the larynx;
         its parts, nervous control of the voice,
         cultivation and care.

IX. *The Special Senses.*

(*a*) The Sense of Sight. Light, refraction, the microscope. The eye as a special organ for receiving impression of light. Structure and function of the eye as to its parts. Defective sight and its remedy. Care of the eyes.

(*b*) Sense of Hearing. Sound, properties of. The ear as a special organ for receiving impressions of sound. Structure and function of the ear as to its parts. Defects of hearing. Care of the ear.

(*c*) The Sense of Touch. Ideas received by touch. The skin papillae organs of touch, where most numerous. Structure and function of papillae. Cultivation of touch.

(*d*) The Sense of Taste. Its nature, value and abuse. Organs of taste. Structure of. Cultivation of taste.

(*e*) The Sense of Smell. Nature of process, value of. Organs of, structure of.

X. *Health and Disease; Poisons and Their Antidotes. Emergencies.*

www.ingramcontent.com/pod-product-compliance
Lightning Source LLC
Chambersburg PA
CBHW020546270326
41927CB00006B/738